SAMPLING PROCEDURES

| Sequential, pp. 26-27 | Stratified, pp. 27-33 | Cluster, pp. 33-37 | Indirect, pp. 38-40 | Area, p. 45 | Capture-recapture pp. 43-44 |

CONFIDENCE INTERVALS FOR:

TYPE OF INTERVAL		Median	Proportion	Mean	Difference between means	Difference between totals	Individuals
	Distribution-free	pp. 55-56	pp. 57-59	pp. 59-61	pp. 61-63	p. 63	pp. 63-65
	Normal curve		p. 66			pp. 66-68	pp. 68-71

Statistical Inference:
The Distribution-free Approach

McGraw-Hill Series in Psychology

Consulting Editors
Norman Garmezy
Richard L. Solomon
Lyle V. Jones
Harold W. Stevenson

Adams *Human Memory*
Beach, Hebb, Morgan, and Nissen *The Neuropsychology of Lashley*
Von Békésy *Experiments in Hearing*
Berkowitz *Aggression: A Social Psychological Analysis*
Berlyne *Conflict, Arousal, and Curiosity*
Blum *Psychoanalytic Theories of Personality*
Brown *The Motivation of Behavior*
Brown and Ghiselli *Scientific Method of Psychology*
Butcher *MMPI Research Developments and Clinical Applications*
Cofer *Verbal Learning and Verbal Behavior*
Cofer and Musgrave *Verbal Behavior and Learning: Problems and Processes*
Crafts, Schneirla, Robinson, and Gilbert *Recent Experiments in Psychology*
Crites *Vocational Psychology*
Davitz *The Communication of Emotional Meaning*
Deese and Hulse *The Psychology of Learning*
Dollard and Miller *Personality and Psychotherapy*
Edgington *Statistical Inference: The Distribution-free Approach*
Ellis *Handbook on Mental Deficiency*
Epstein *Varieties of Perceptual Learning*
Ferguson *Statistical Analysis in Psychology and Education*
Forgus *Perception: The Basic Process in Cognitive Development*
Franks *Behavior Therapy: Appraisal and Status*
Ghiselli *Theory of Psychological Measurement*
Ghiselli and Brown *Personnel and Industrial Psychology*
Gilmer *Industrial Psychology*
Gray *Psychology Applied to Human Affairs*
Guilford *Fundamental Statistics in Psychology and Education*
Guilford *The Nature of Human Intelligence*
Guilford *Personality*
Guilford *Psychometric Methods*
Guion *Personnel Testing*
Haire *Psychology in Management*
Hirsch *Behavior-genetic Analysis*
Hirsh *The Measurement of Hearing*
Hurlock *Adolescent Development*
Hurlock *Child Development*
Hurlock *Developmental Psychology*
Jackson and Messick *Problems in Human Assessment*
Karn and Gilmer *Readings in Industrial and Business Psychology*
Krech, Crutchfield, and Ballachey *Individual in Society*
Lazarus *Patterns of Adjustment and Human Effectiveness*
Lazarus *Psychological Stress and the Coping Process*
Lewin *A Dynamic Theory of Personality*
Lewin *Principles of Topological Psychology*
Maher *Principles of Psychopathology*
Marx and Hillix *Systems and Theories in Psychology*
Messick and Brayfield *Decision and Choice: Contributions of Sidney Siegel*
Miller *Language and Communication*
Morgan *Physiological Psychology*
Nunnally *Psychometric Theory*

Rethlingshafer *Motivation as Related to Personality*
Robinson and Robinson *The Mentally Retarded Child*
Scherer and Wertheimer *A Psycholinguistic Experiment on Foreign Language Teaching*
Shaw and Wright *Scales for the Measurement of Attitudes*
Sidowski *Experimental Methods and Instrumentation in Psychology*
Siegel *Nonparametric Statistics for the Behavioral Sciences*
Stagner *Psychology of Personality*
Townsend *Introduction to Experimental Methods for Psychology and the Social Sciences*
Vinacke *The Psychology of Thinking*
Wallen *Clinical Psychology: The Study of Persons*
Warren and Aker *The Frontal Granular Cortex and Behavior*
Waters, Rethlingshafer, and Caldwell *Principles of Comparative Psychology*
Winer *Statistical Principles in Experimental Design*
Zubek and Solberg *Human Development*

John F. Dashiell was Consulting Editor of this series from its inception in 1931 until January 1, 1950. Clifford T. Morgan was Consulting Editor of this series from January 1, 1950 until January 1, 1959. Harry F. Harlow assumed the duties of Consulting Editor from 1959 to 1965. In 1965 a Board of Consulting Editors was established according to areas of interest. The current board members are Richard L. Solomon (physiological, experimental), Norman Garmezy (abnormal, clinical), Harold W. Stevenson (child, adolescent, human development) and Lyle V. Jones (statistical, quantitative).

Statistical Inference: The Distribution-free Approach

Eugene S. Edgington

Professor, Department of Psychology
University of Calgary

McGraw-Hill Book Company

New York
St. Louis
San Francisco
London
Sydney
Toronto
Mexico
Panama

To My
Mother and Father

Statistical Inference: The Distribution-free Approach

Copyright © 1969 by McGraw-Hill, Inc. All rights reserved. Printed in the United States of America. No part of this publication may be reproduced, stored in a retrieval system, or transmitted, in any form or by any means, electronic, mechanical, photocopying, recording, or otherwise, without the prior written permission of the publisher.

Library of Congress Catalog Card Number 74-83265

18979

1234567890   MAMM   7654321069

Preface

The objective of this book is to provide an approach that will be useful in working toward a complete understanding of the fundamentals of statistical inference. This approach is called "distribution-free" because it concerns finite, discrete populations of unknown shape. The distribution-free approach to statistical inference is simply to think about statistical inference in terms of such populations. This approach permits a penetrating analysis of statistical inference without recourse to calculus or other advanced mathematical techniques needed for a comparable analysis in terms of infinite, continuous distributions, like the normal curve.

In order to provide a foundation upon which to build a system of statistical inference for finite populations, it was necessary to define probability and random sampling strictly in terms of finite populations and to revise a number of concepts accordingly. These preliminary problems are treated in Chapters 2 and 3. In addition to the discussion needed to clarify the concept of random sampling of finite, discrete populations, Chapter 3 includes a discussion of various sampling procedures not usually treated in statistics books.

Chapter 4 concerns an aspect of distribution-free methods that may be unfamiliar to many persons: distribution-free procedures for constructing confidence intervals for population parameters. These procedures in general provide much less precise confidence intervals than normal curve procedures, and so their practical value is limited, but they are particularly useful for providing insight into the assumptions underlying both distribution-free and normal curve confidence intervals.

Chapter 5 deals with the application of randomization tests for testing hypotheses about experimental treatment effects. Analysis of the problems of drawing inferences about treatment effects leads to views that differ considerably from traditional views on the design of experiments and statistical analysis of experimental data.

Chapter 6 extends the distribution-free approach. The role of regression procedures in experimentation is examined. A distribution-free procedure for detecting nonmonotonic relationships is presented.

Chapter 7 illustrates the application of the distribution-free approach to a complex problem: joint analysis of directions of difference in central tendency and variability. A brief discussion is also given of the interpretation of a special class of interaction.

Many ideas in this book will be of general interest not simply because they are new but because they are in direct opposition to commonly held beliefs about statistical inference. The principal importance of these ideas, however, is that their development in the book illustrates both the method and the value of the distribution-free approach.

Lyle Jones read the manuscript at various stages in its development and was especially helpful in indicating where the logic needed to be clarified.

Financial aid from the University of Calgary research fund is gratefully acknowledged.

Eugene S. Edgington

Contents

Preface	vii
Chapter One **Overview**	1
The distribution-free approach	2
Logical relationships between statistical models and inferences about the world of application	2
The organization of this book	3
Chapter Two **Probability**	7
Three concepts of probability	7
An operational definition of probability	9
The need for formal probability models	11
Dice	11
Coins	13
Maturing of the chances	15
Uses of probabilities	16
Probability models for distribution-free interval estimation	16
Probability models for randomization tests	17
Chapter Three **Sampling**	21
Random sampling	21
Measurements are not randomly sampled	23
Reasons for not studying entire populations	23
Analysis of variability: the key to efficient sampling	24
Random sampling techniques	26
Sequential sampling	26
Stratified sampling	27
Cluster sampling	33
Sources of variability	37
Indirect sampling	38
Sampling subpopulations	40
Sampling to estimate the size of animal populations	42
Area sampling	45
Bias from sample dropouts	45
Haphazard sampling	47
Chapter Four **Distribution-free Confidence Intervals**	49
Purposes of interval estimation	50
Distribution-free confidence intervals	53
Distribution-free confidence interval for a median	55
Distribution-free confidence interval for a proportion	57
Distribution-free confidence interval for a mean	59
Distribution-free confidence interval for a difference between means	61

Distribution-free confidence interval for an individual	63
The relative efficiency of distribution-free estimators for the mean and the median	65
Normal curve confidence interval for a proportion	66
Normal curve confidence interval for a difference between totals	66
Normal curve confidence interval for an individual	68
Confidence interval for an individual based on linear-regression procedures	71
The normal distribution assumption	73
The central limit theorem argument for normality	73
The binomial distribution argument for normality	76
Nonmathematical arguments for and against normality	77
Transformations and normality	79
The bivariate normal distribution assumption	81
Validity of measurements	83
Reliability of measurements	84
Inferences about measurable individuals	87
Statistical and nonstatistical inferences	89
Estimation in psychological testing	90

Chapter Five **Randomization Tests for Experiments** 93

Random assignment and randomization tests	93
Levels of significance and probability	95
The null hypothesis for randomization tests	95
The widespread use of nonrandom samples in psychological experiments	96
The irrelevance of random sampling for psychological experiments	97
Randomization test for a difference between independent samples	98
Randomization test for paired comparisons	105
Randomization test for contingency	109
Randomization test for correlation	115
Randomization test for interaction	118
Other randomization tests	120
Rank-order tests	123
Rank-order test for independent samples	123
Rank-order test for paired comparisons	124
Contingency chi-square test	125
Rank correlation	128
Tied ranks	130
One-subject experiments	135
Aspects of experimental design to facilitate nonstatistical generalization	140
The design of longitudinal experiments	142
Statistical independence	146
Experimental independence	150
Approximate randomization tests	152
Transformations	155
Normal curve tests as approximations to randomization tests	161
Analysis of variance	162

t test for paired comparisons	164
Product-moment correlation	166
Must null hypotheses be false?	167

Chapter Six **Experimental Regression** — 169

The objectives of experimental regression	169
General laws	170
Aspects of design to facilitate nonstatistical generalization	171
Fitting regression lines to data	172
Spacing of the independent variable	173
Estimating the simultaneous-measurement regression line	173
Estimating the perfect-experimental-control regression line	175
Implications of symmetry about a regression line	176
Randomization test for a difference between regression lines	177
A statistical test for nonmonotonic trends	178

Chapter Seven **Joint Analysis of Differences in Central Tendency and Variability** — 183

Comparisons of frequency distributions	183
Types of differences in central tendency and variability	184
Type *A* situation	186
Type *B* situation	186
Type *C* situation	187
Type *D* situation	188
Type *E* situation	189
The failure to randomly assign individuals to high-scoring and low-scoring groups	190
A distribution-free analog of a complex analysis of variance design	191
Interactions between treatments and measurement magnitudes	197
The effect of transformations on interactions	198

References	203
Index	207

Chapter One

Overview

It is hoped that this book will show how applied statistics can be developed more logically. There is no doubt, of course, that the mathematics associated with applied statistics is rigorous, but the nonmathematical reasoning seems at times not to be strictly logical. For example, are experimenters logically justified in drawing statistical inferences about other populations than those from which they have taken random samples? In what sense can the recorded responses of an experimental subject be regarded as a random sample of his responses? Can the practice of carefully constructing test items, trying them out, and keeping the best items be regarded as a procedure of random sampling of a population of test items? And are we justified in making statistical inferences about the future success of students, when in fact we cannot randomly sample the future?

Each of the preceding examples concerns random sampling. This is characteristic of logical problems in the area of applied statistics. An understanding of the nature of random sampling and its role in statistical inference should, then, help one to reason about applied statistics in a more logical manner.

It appears that in our statistics courses, by emphasizing the mathematical properties of random sampling we have neglected to put sufficient stress on the construction of physical procedures to be used in random sampling. In this book the lottery is regarded as the ideal physical procedure for random sampling. It will be shown that there are numerous advantages in thinking about random sampling in terms of lotteries, not the least of which is that this forces one to recognize the finite and discrete character of any population which can in practice be randomly sampled.

In order to provide a proper framework for a discussion of physical procedures for random sampling, and for lottery sampling in particular, it will be necessary to revise certain definitions and concepts of probability and sampling. More than that, we must face the more basic problem: the lack of understanding of the relationship between statistical models and statistical inferences about the empirical world. It is necessary to carefully determine what the world of statistics and the world of application have in common and this must be done in a way that is comprehensible to nonmathematicians, since most persons who apply statistics are not mathematicians.

Thus, in attempting to show how applied statistics can be developed more logically, the principal method will be to carefully examine the link between abstract statistical models and the concrete physical events to which they are applied.

The distribution-free approach

This book demonstrates the distribution-free approach to statistical inference. The distribution-free approach is to think about statistical inference in terms of finite discrete populations of unknown shape. The principal objective of the distribution-free approach is to gain a better understanding of the logical connections between statistical models and the world of application.

The reason for basing the distribution-free approach on finite discrete populations is that populations in the world in which statistics is applied *are* finite and discrete. An important consequence of referring statistical inference to such populations is that it permits the nonmathematician to gain insight into the fundamentals of statistical inference because finite discrete populations, by virtue of their freedom from the concepts of infinity and continuity, can be analyzed without the necessity for advanced mathematics.

The distribution-free approach regards populations as having unknown shapes for the same reason that it regards them as being finite and discrete: this is the nature of populations in the world of applied statistics. In practice, we may *guess* the general shape of the population being sampled, but we never *know* its shape, and so it is appropriate to use an approach that concerns populations of unknown shape. We can then examine statistical inference in a setting as free as possible from dubious empirical assumptions.

Logical relationships between statistical models and inferences about the world of application

Statistical procedures have developed from a need for dealing with particular types of problems in the empirical world. Nevertheless, textbooks commonly place much more emphasis on the mathematical relationships between various aspects of a statistical model or between different statistical models than on the logical relationships between the models and inferences about the physical world based on application of the models. For example, there is a tendency in statistics books to stress algebraic derivations of formulas and the algebraic equivalence of alternative formulas. Sometimes four or five formulas are given for the product-moment correlation coefficient, all of them mathematically equivalent. Yet consideration of the link between statistical models and the empirical world is usually restricted to showing how to substitute measurement numbers into formulas. Study of the mathematics of statistics leads to facility in that aspect of statistics, but there is no evidence that it leads to facility in other aspects. A person may have a thorough understanding of the mathematics of statistics without having the slightest notion of the proper application of statistical models to empirical problems.

The simplicity of the distribution-free procedures makes them ideal for showing the relationship between statistical models and inferences about empirical processes. In numerous examples, distribution-free procedures will be developed to solve a particular empirical problem. By starting with an investigator's problem and developing a statistical procedure especially for that type of problem, it is possible to show quite clearly the logical connections between the statistical models and the inferences based on them. The justification for probability statements thereby becomes obvious. These are not simply examples to show how to apply a statistical technique; they show the development of the technique, and they show it in sufficient detail to provide the reader with an understanding of the general principles of the derivation. Few persons will have difficulty generalizing from such examples to quite different situations.

This book illustrates the usefulness of the distribution-free approach by applying it to an analysis of the relationship between statistical models and their application. This procedure not only casts new light on already known facts, but also leads to the discovery of new facts. The distribution-free approach has both pedagogical and heuristic value.

The organization of this book

The organization of this book can best be understood in terms of its objective: to illustrate the application of a special way of thinking about statistical inference. The reader who loses sight of this objective may waste time looking for a type of organization that was never intended.

We will now briefly consider the objectives of the individual chapters and their organization.

4   Statistical Inference: The Distribution-free Approach

Chapter 2: Probability

This is a short chapter which has the principal objective of examining the relationship between probability models and probability applications. It considers the purpose of probability models to be to aid people in deriving the logical consequences of certain assumptions. In order to obtain probability values of general interest, the computations should be based on the application of physical procedures like the lottery or certain games of chance where there is general agreement regarding equally likely events. This is the reason lottery sampling is regarded in Chapter 3 as the basic sampling procedure.

Chapter 3: Sampling

This chapter applies the concept of lottery sampling to various types of problems, simple and complex. Special types of sampling, like stratified and cluster sampling, are considered. It is demonstrated that continuous populations such as geographical areas can be sampled by means of the lottery after they have been subdivided into a finite population of units. Sampling procedures unlike the lottery are examined to see whether they are biased procedures.

Chapter 4: Distribution-free confidence intervals

This chapter concerns confidence-interval estimation for finite, discrete populations of unknown shape. Since all population elements are represented (by pellets or some other object) in a lottery, there is obviously no problem in determining the population size, and so the size of the population is assumed to be known. The distribution-free confidence intervals in general are much less precise than normal curve confidence intervals for the same parameter and therefore have little *practical* value. They do have *theoretical* importance however because they lead to a better understanding of the assumptions underlying both distribution-free and normal curve confidence-interval estimation. In the process of developing distribution-free confidence-interval estimation procedures, the possibility of analogous normal curve procedures arises. For instance, in constructing a distribution-free confidence interval for the mean, it is necessary to essentially construct a confidence interval for the individuals in the population that are not in the sample. Thus, it is shown that confidence intervals can be constructed not just for population parameters but also for other samples from the same population. This suggests the possibility of a similar procedure for normal curve estimation, which is found to be derivable from existing formulas with very little mathematical manipulation.

Chapter 5: Randomization tests for experiments

This chapter logically depends on the probability and sampling chapters, just as Chapter 4 does, but it does not depend on Chapter 4. This chapter, unlike the previous one, demonstrates distribution-free procedures that are not only of theoretical importance but of practical usefulness, as well. It was unnecessary to devise special hypothesis-testing procedures for finite discrete populations of unknown shape because they already exist in the form of randomization tests. Randomization tests of various kinds are discussed and the corresponding rank-order statistical tests are described. In exploring the relevance of random sampling of finite populations to psychological experiments, it is found that seldom are such experiments intended for the purpose of drawing inferences about a specific existing population. Therefore, seldom is random sampling relevant to the objectives of the experiment. Random assignment rather than random sampling is the appropriate procedure for introducing the random element necessary to permit statistical inferences. The concept of random sampling seems to have little relevance to psychological experimentation. Statistical inferences about experimental subjects can be made on the basis of random assignment, and inferences about other individuals must be nonstatistical and therefore be based on empirical rather than probabilistic arguments. Normal curve statistical tests are regarded as approximations to randomization tests.

Chapter 6: Experimental regression

This chapter deals with the use of regression procedures in experiments. The concepts associated with the use of regression procedures in *nonexperimental* settings are found to be inappropriate. A distribution-free procedure for detecting both monotonic and nonmonotonic correlations and the probability table for it are presented. It is found that statistical inferences about experimental regression are quite narrow, and so consideration is given to nonstatistical inferences and the types of arguments that are needed to support them.

Chapter 7: Joint analysis of differences in central tendency and variability

The link between statistical models and empirical problems is carefully studied in this chapter in regard to a complex problem: how to use, in a comparison of two groups, the relationship between the direction of difference in central tendency and the direction of difference in variability to draw inferences about the underlying empirical processes. After treating the problem independently of distribution-free procedures, the problem is reexamined in the light of randomization-test considerations. This leads to a randomization test associated with a more complex experimental design than those in Chapter 5.

Chapter Two

Probability

Statistical inference requires the use of probability, and so an understanding of probability is fundamental to an understanding of statistical inference. A clear understanding of statistical inference cannot be based on a fuzzy understanding of probability. It is essential, therefore, that special consideration be given in this chapter to the most perplexing aspect of probability: the relationship between a probability model and empirical inferences based on the model. A proper appreciation of the rest of the book depends on apprehending the nature of this relationship.

Three concepts of probability

Some persons use the word "probability" to refer to their strength of belief in something; they say that there is a high probability of an event occurring if they have a strong belief that it will occur. For instance, a person will say that there is a high probability that a particular horse will win a race if he strongly believes that the horse will win. *Subjective probability*, as this kind of probability is sometimes called, has been the object of considerable investigation but no system of statistical inference based on it has been generally accepted by statisticians. The lack of wide acceptance of subjective probability is in part a consequence of the limited scientific use for the numerical probability values, which are based on personal knowledge and belief.

A second concept of probability is the *long-run relative frequency* concept. According to this view of probability, the probability of an event is the relative frequency of occurrence of the event in the long run. In other words, the probability of an event occurring is the proportion of the time the event occurs out of all occasions for which the occurrence or nonoccurrence of the event is to be considered, for infinitely many occasions. For instance, for an unbiased die, the probability of rolling a 4 is said to be 1/6 because it is assumed that, in the long run, one-sixth of the rolls will give a 4. The major flaw in this definition of probability is that the expression "in the long run" refers to an infinite series of events whereas no physical object, whether it be a die, a coin, or anything else, can be subjected to infinitely many manipulations for which the outcome can be observed. A die or a coin can be tossed only a finite number of times before it wears out, and people can keep records of the outcome of only a finite number of tosses; in both cases, the number of tosses may be large, but it is finite. Because relative frequency in the long run refers to an infinite series, any numerical probability value referring to such a relative frequency (proportion) is unverifiable. The unverifiability of such probabilities may be the reason statements regarding empirical relative frequencies "in the long run" so often go unchallenged. After all, it is easier to believe in a magical *infinite* series that compensates for deviations from some specified relative frequency than to believe that the specified relative frequency of an event is the exact proportion of trials on which the event will occur in a specified *finite* series of trials. People are much more inclined to accept the notion that in the long run a coin will show up heads exactly half of the time than they are to believe that in 5 million tosses it will show up heads exactly half of the time. As soon as a limit on the number of tosses is set, no matter how high, it seems less plausible that the ratio of heads to tails will be exactly 1:1. The definition of probability as the relative frequency in the long run is certainly not an operational definition, because there are no possible empirical operations for determining such a probability. As long-run relative frequencies can never be observed they necessarily will be unknown and cannot be the basis for assigning numerical probability values. What appear to be long-run relative frequency probability values, such as the probability of 1/2 for tossing a head with an unbiased coin, can be regarded as *estimates* of long-run relative frequencies, based on alternative considerations, as for example considering that on any given toss there is as much reason to expect a head as a tail.

The *classical* definition of probability of occurrence of an event is the number of equally possible ways in which that event can occur divided by the number of equally possible ways any event can occur. To illustrate, the probability of tossing a head with an unbiased coin is 1/2 because there is one way to get a head out of two equally possible ways of getting any event; i.e., tossing a head or tossing a tail. There are six sides on a die and two sides have numbers less than 3, and so the probability is 2/6 of throwing a number less than 3. Although similar to the long-run relative frequency concept of probability, the classical concept is different in important respects. The intent of the classical definition is to prescribe a computational procedure for exact determination of a numerical probability value, whereas the long-run relative frequency definition does not specify how to determine the numerical probability value. In order to compute probabilities by following the classical definition, however, it is necessary to determine which events are equally possible. How this is to be determined is usually left unanswered. If "equally possible" refers to equal long-run relative frequencies, the classical definition of probability is no better than the long-run relative frequency definition, because the long-run relative frequencies cannot be determined. If, on the other hand, "equally possible" refers to some objective empirical property, it is unclear what sort of examination would provide evidence that events were equally possible. The expression "equally possible" when considered to have objective reference has the same sort of ambiguity as that associated with the term "unbiased" as applied to a coin. The impression that the word "unbiased" conveys is that there is some objective property corresponding to it, such as the symmetrical shape and uniform density of the coin. It is easy to forget that the symmetry of a coin or die is not the only thing that affects the outcome, easy to overlook the necessity for imposing special restrictions on the tossing procedure to ensure that it is a "fair toss." In attempting to describe an unbiased tossing procedure, therefore, it becomes clear that terms like "equally possible," "unbiased," and "equal chance" do not refer to objective empirical properties at all but refer instead to subjective attitudes or judgments that reflect the anticipated outcome of a trial or series of trials.

An operational definition of probability

An *operational definition* of probability will be attempted here, based on the way statisticians compute probability values instead of the way they define probability. Judging by the way statisticians compute probabilities, the probability of a type A event within a class of events is the fraction of the events which are type A events. This computational procedure settles only part of the problem. For what situations should probabilities be computed in this manner? The selection of appropriate situations for such computations is a subjective matter; the computation can be carried out in the above manner for situations where all events are *equally likely* events. The term "equally likely" is used in preference to "equally possible" to emphasize the *subjective* basis of the designation.

Two or more events are said to be equally likely for a person if he considers it just as reasonable to expect one event as another. Also, a person may class events as equally likely if he expects one to occur as often as another in the long run. (A person may have some sort of notion, no matter how vague, of relative frequency in the long run, even though there is no empirical way to determine such a relative frequency.) If situations did not have to meet further requirements to permit the computation of a probability value, the computed probabilities would be highly personal in many cases, and of little interest to other persons, because events that one person would judge to be equally likely would not be judged so by other persons. We will next consider what further requirements should be imposed on the computation of a probability value in order to ensure that it will be generally acceptable.

In the desire for probability values that would have wider acceptance from person to person many statisticians have implicitly imposed the condition of *sufficient reason*. This means that events must be judged equally likely only if there is sufficient reason to expect one event to occur as often as another. Rather than specify exactly what is meant by "sufficient reason," it is customary to contrast it with *insufficient reason*, which is based on ignorance. A person using "insufficient reason" could suppose that a head is as likely as a tail for a toss of a particular coin that he has never examined, simply because he has no reason to suppose otherwise. That this is not a desirable way to determine equally likely events is obvious because the coin may have a tail on both sides. According to the principle of sufficient reason a person should use the concept of equal likelihood only with regard to situations in which knowledge rather than ignorance leads to the designation of events as equally likely. Modern statisticians, in their desire to have unanimity among persons regarding equal likelihoods in a given situation, want equal likelihoods to be based on carefully constructed procedures such as lotteries, or carefully controlled games of chance. This last requirement, that there be agreement among persons regarding equality of likelihoods, would not be imposed by all statisticians; certainly Bayesian statisticians would not make this requirement. However, the conventional conservative statistician would agree with making this requirement, although he might never explicitly state it.

To sum up, there are two parts to the procedure generally accepted by statisticians for providing probabilities. First, there must be a generally acceptable procedure for assigning equal likelihoods to events. Second, for a general class of equally likely events the probability of getting an event within a particular subclass is determined by finding the fraction of the events that fall within that subclass. For example, with careful construction of a die and control over tossing, there would be general agreement on the equal likelihood of the six possible outcomes. For instance, the probability of either a 2 or a 3 would be computed as 2/6 since 2/6 of the equally likely events are in the class "2 or 3."

This operational definition involves both an objective aspect, which is the computation, and a subjective aspect, which is the determination of equal likelihood. The subjective aspect is made more palatable by requiring that equal likelihoods be determined only by procedures that are generally acceptable, thereby endowing probabilities with public rather than private significance.

The need for formal probability models

Like other models, a probability model is a device to help a person derive the logical consequences of certain postulates or assumptions. The assumptions a person makes determine which probability model he should use. It could be said that a person introduces into a probability model certain characteristics of the way he thinks about things.

The probability of a simple event in a class of events can be determined directly without the aid of a formal model, but in more complicated cases, models help a person visualize the situation. Even capable mathematicians sometimes need schematic diagrams to guide them in the computation of probabilities. The type of schematic diagram which is used in this chapter is called a *probability tree*.

Dice

Example 2.1

Smith decides to investigate probability to help him recognize a fair bet in gambling. In particular, he is interested in the odds with dice. Nobody that he knows gambles with just one of a pair of dice, but he believes the way to understand the odds is to start with the roll of a single die and then, after the probabilities for a single die are understood, to study a pair of dice. Smith examines a die and is convinced that it is not loaded, shaved, or otherwise "unfair." Furthermore, he decides to use a dice cup and a board from which to bounce the die to ensure fairness of the toss. By virtue of these considerations, Smith considers all six sides equally likely to turn up. The 4 is no more to be expected than the 6, the 5 no more than the 3, and so on. Under standard dice-rolling conditions, Smith would expect one of the sides of the die to show up about as often as any other side. He represents his notion of equally likely events in a simple probability model in which each of the events considered equally likely (numbers on the die) is represented by one element in the model on the next page:

12 Statistical Inference: The Distribution-free Approach

1 2 3 4 5 6

With these six numbers representing the corresponding sides of the die, this probability model can be used to determine probabilities which Smith can convert to fair betting odds. The probability of rolling an even number is seen to be 3/6 because three of the six equally likely events represented are even numbers. The probability of rolling an odd number is, of course, also 3/6. Equal probabilities of two events implies equal betting odds. Thus, since the probability of an even number is the same as that of an odd number, Smith decides that 1:1 is the proper odds for rolling an even number. The probability of getting either a 4 or a 5 is 2/6 because there are two events of the six that are members of the class "4 or 5." Since the other four of the six are *not* members of the class "4 or 5" the betting odds in favor of rolling a 4 or a 5 are 2:4. Alternatively, the odds *against* rolling a 4 or a 5 are 4:2.

Example 2.2

Having mastered the art of determining odds for the toss of a single die, Smith decides to move on to a more practical problem: determining the odds for a pair of dice. He wants to extend the model for a single die to represent the way he conceives of a toss of a pair of dice. He believes that under controlled dice-throwing conditions, it is reasonable to conceive of the outcome of the toss of two dice as equivalent to the outcome of two successive throws of a single die in which neither the number of previous tosses nor the particular outcome on any previous toss has any influence on the outcome of the next toss to be made. Therefore, the concept of equal likelihood of each of the sides of a die is unaffected by the results of previous rolls and similarly, the concept of equal likelihood of each side of one die is unaffected by the result of the roll of the other die when two dice are tossed. This concept is represented in this way: for each of the six possible results of the toss of one die list six possible results of the toss of the other die. This is done in the following diagram, in which one of the dice is arbitrarily called die *A* and the other, die *B*:

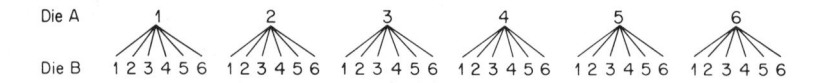

```
Die A      1           2           3           4           5           6
          /|\         /|\         /|\         /|\         /|\         /|\
Die B  1 2 3 4 5 6  1 2 3 4 5 6  1 2 3 4 5 6  1 2 3 4 5 6  1 2 3 4 5 6  1 2 3 4 5 6
```

The lines connecting numbers in the first row with numbers in the second row represent possible results: a 1 on die *A* and a 1 on die *B*; a 1 on die *A* and a 2 on die *B*, etc. There are 36 lines representing 36 possible results for the two dice. Each of these 36 results is equally probable, having a probability of 1/36. Here there are classes and subclasses of *pairs of numbers* in the model rather than classes and subclasses of *numbers*. In this model the probability of rolling a total of 6 with two dice is 5/36, because five of the 36 lines connect numbers that add up to 6: 1-5, 2-4, 3-3, 4-2, and 5-1. There is only one way to roll two 3s, whereas there are two ways to roll a 2 and a 4, and so the probability of rolling two 3s is 1/36, while the probability of rolling a 2 and a 4 is 2/36. These examples illustrate how a probability model can aid a person in deriving the logical consequences of certain assumptions. The assumption of equal likelihood of all sides for a single die, together with the assumption of independence of the outcomes on the two dice, implies that each pair of the 36 pairs of numbers is to be considered equally likely, because each pair of connected numbers is represented once in the model.

Coins

Example 2.3

The value of having a systematic procedure for representing repetitions of events cannot be exaggerated. To take an example, Jean D'Alembert, a prominent eighteenth-century mathematician, stated that for independent tosses of an unbiased coin, the probability of two heads in two tosses is 1/3. Implicitly, his "model" was a model of two tosses, showing the outcomes in this manner:

two heads   head and tail   two tails

Actually, the appropriate model for two independent tosses of an unbiased coin is the following:

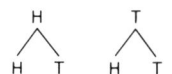

```
     H           T
    / \         / \
   H   T       H   T
```

This probability model shows that the probability of two heads in two tosses is *1/4*. By not representing the outcomes of the first toss and those of the second toss separately, D'Alembert did not see the need for distinguishing between T-H and H-T, and thus failed to see that *each* of these events should be considered as likely as H-H (or as likely as T-T).

Example 2.4

Jones is going to bet Thomas that Thomas will not get as many as two heads in three tosses of a coin. To determine the correct odds, he

14 Statistical Inference: The Distribution-free Approach

considers the tosses to be independent and represents the equally likely outcomes of three tosses in this way:

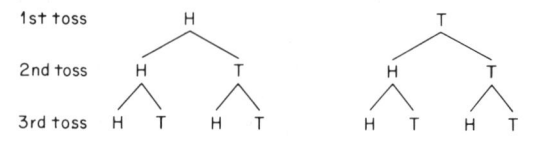

```
1st toss              H                         T
                     / \                       / \
2nd toss            H   T                     H   T
                   /\   /\                   /\   /\
3rd toss          H  T H  T                 H  T H  T
```

Four of the eight equally probable outcomes of three tosses of a coin do not involve as many as two heads, and so the probability is 4/8 that there will not be as many as two heads in three tosses. Therefore, Jones bets $5 against Thomas's $5 that Thomas will not get as many as two heads in three tosses.

Example 2.5

In Example 2.4 Jones bet $5 against Thomas's $5 that Thomas would not throw at least two heads in three tosses of a coin. After Thomas has made two tosses and thrown two heads, a third person, Harris, joins them. He is told about the bet between Jones and Thomas and is told that Thomas has thrown two heads in the first two tosses. He is asked if, before the final toss is made, he wants to make a bet that Thomas will not have at least two heads in three tosses. Thomas eagerly points out that the probability model for three tosses (shown in Example 2.4) shows that the correct odds are 1:1 against as many as two heads in three tosses, and suggests that Harris could make the same bet as Jones: $5 against Thomas's $5 that there will not be as many as two heads after the third toss. Harris knows that Thomas would win because he has already thrown two heads, but he wonders why the probability model suggests odds that contradict good judgment. He decides that the probability model that was appropriate for Jones was one that represented a way of thinking about equally likely events before *any* tosses were made, but that he himself needs a model for conceiving of equally likely events after the first two tosses have already yielded heads. This model would be:

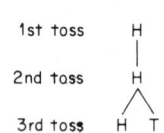

```
1st toss        H
                |
2nd toss        H
               / \
3rd toss      H   T
```

In this model, the probability of at least two heads is 1, and so it would be foolish to bet against getting at least two heads. However, the probability of three heads is 1/2, and so Harris decides to bet Thomas $5 against $5 that he will not have *three* heads after the next toss.

Maturing of the chances

Some betting systems, or *martingales* as they are sometimes called, are implicitly based on the notion of *interdependence* of trials in situations where the gamblers would explicitly state that they considered each trial to be *physically independent* of previous trials. To take a common example, after a coin has been tossed several times and has turned up heads each time, some persons believe that the coin is somewhat more likely to turn up tails on the next toss. Some of these persons use reasoning that presupposes a long-run relative frequency of 1/2 for heads. Assuming that up to that time the coin has had slightly more than 1/2 heads, these persons believe that the excess of heads must be compensated for by an excess of tails in the future. This type of imagined compensation has been dignified by the label "maturing of the chances." These persons persist in this belief even when they are convinced that there is no *physical* connection between one toss of a coin and the next toss capable of producing this effect. They believe in a type of *statistical* interdependence which operates independently of physical causes. As they put it, the *law of averages* exercises control over the tosses. Apparently this strange belief stems from faith in the ability of a long-run relative frequency probability to exert a retroactive effect on individual tosses. Admittedly, many people who subscribe to the long-run relative frequency concept of probability do not believe in a law of averages that works in this way, but there is little doubt that the mystery that surrounds an *infinite* (i.e., long-run) series stimulates belief in a purely statistical determinism which operates independently of physical laws.

It is possible that some believers in the "maturing of the chances" are not basing their belief on the long-run relative frequency concept but are simply using a good probability model in the wrong context. The following example shows how this might occur.

Example 2.6

There are situations where it is appropriate to think in terms of "maturing of the chances." Consider as an instance the model for taking random samples of two elements, without replacement, from a finite population of five elements: 1, 2, 3, 4, and 5. Represent equally likely samples by the diagram on the following page:

16 Statistical Inference: The Distribution-free Approach

```
1st draw      1           2           3           4           5
             /|\         /|\         /|\         /|\         /|\
2nd draw   2 3 4 5     1 3 4 5     1 2 4 5     1 2 3 5     1 2 3 4
```

In this model, equal likelihoods based on random sampling are the justification for equal representation of all possible outcomes. In sampling without replacement, the element that has been drawn cannot be drawn again, and so the trials are *not* independent. Nonetheless, each of those elements *not* drawn on the first draw is considered equally likely to be drawn on the second draw. Consider the class of odd numbers 1, 3, and 5. Suppose that in the first draw a 5 was drawn. Then the probability of getting a 1, a 3, or a 5 on the second draw is not 3/5 as it was on the first draw but is 2/4, because two of the four remaining elements are in that class. On the other hand, if the first number drawn was not in that class, but was, for example, a 4, the probability is then 3/4 that the next number drawn will be in the class of odd numbers. In other words, when a population is sampled without replacement, the elements that are drawn are "used up," thereby reducing the probability of drawing an element from the same class on the next draw. (This is true only when there are elements of a different class in the population at the time of the draw. For example, if both even numbers have been drawn, and none of the odd numbers, drawing an odd number on the third draw will not, of course, reduce the probability of drawing an odd number on the fourth draw.) In sampling without replacement, then, the notion of "maturing of the chances" is frequently appropriate. The model for sampling without replacement is obviously *not* the appropriate model, however, for dice, roulette, or other situations where the trials *are* independent.

Uses of probabilities

In the preceding examples determining the probabilities of different events was for the purpose of recognizing fair odds for betting. That is one use of probabilities. Another is to serve as an estimate of the relative frequency with which certain events occur. A third use of a probability is to indicate for a future occasion which events have high probabilities in order that a person may plan for such events. And, as was stated at the beginning of this chapter, probability is the basis of statistical inference.

Probability models for distribution-free interval estimation

In the chapter on confidence intervals the probability values will be computed without direct reference to probability tree diagrams because the diagrams would be unduly large. The computed probability values, however, are precisely those that would be obtained by the use of a probability tree.

In distribution-free estimation it is often useful to consider equally likely groups of scores *not* in the sample; i.e., those scores remaining in the population after a sample has been drawn.

Example 2.7

A random sample of three scores is taken without replacement from a population of five scores. Assume that all five scores are different and that we represent them by the rank numbers 1, 2, 3, 4, and 5, where rank 1 is the largest score. What is the probability that one of the two scores *not* in the sample is the largest score in the population? This probability is precisely the same as the probability that one of two scores *randomly selected* is the largest score because two scores left in random sampling can be regarded as two scores *randomly selected to remain*. Thus, we can use the following diagram of equally likely pairs of remaining scores to determine our answer:

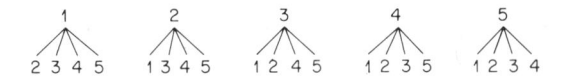

```
     1         2         3         4         5
    /|\       /|\       /|\       /|\       /|\
   2 3 4 5   1 3 4 5   1 2 4 5   1 2 3 5   1 2 3 4
```

Out of twenty equally likely pairs of *undrawn* scores, eight contain a 1, and so the probability is 8/20 that one of the last two scores is the largest score in the population.

Probability models for randomization tests

As in the chapter on confidence intervals, the probability values in the chapter on randomization tests will be computed without the use of probability tree diagrams. It is frequently desirable, however, to visualize a probability tree to see precisely how to compute the probability for a randomization test.

Example 2.8

Take the problem of determining the probability that a random pairing of two sets of scores will provide a perfect positive rank correlation. This is a permutation problem. Let one set of scores consist of 5, 8, and 10 and the other set be 1, 4, and 7. By random pairing is meant a procedure devised to make every possible pairing of the two sets of scores seem equally likely.

18 Statistical Inference: The Distribution-free Approach

For instance, such a procedure would be to put the numbers in the second set in a hat and to draw one of the numbers and pair it with the 5, draw another number and pair it with the 8, and draw the last number and pair it with the 10. In the following diagram we represent in the first line the equally likely pairings with score 5, in the second line the equally likely pairings of score 8 with the remaining scores in the second set, and in the third line the pairing of 10 with the last number:

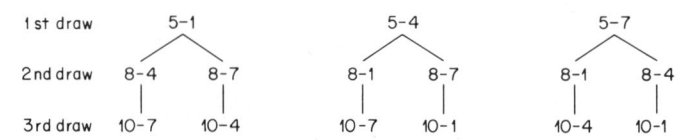

```
1st draw        5-1              5-4              5-7
               /  \             /  \             /  \
2nd draw    8-4    8-7       8-1    8-7       8-1    8-4
             |      |         |      |         |      |
3rd draw   10-7   10-4      10-7   10-1      10-4   10-1
```

We have here a probability model showing six equally likely ways of pairing the two sets of scores, each way represented by one of the branches connecting the three rows. A perfect positive rank correlation means that 5, 8, and 10 are paired in that order with 1, 4, and 7. Thus, a perfect positive rank correlation is represented in the model by the branch 5-1, 8-4, 10-7. This branch is one out of six possible branches, and so the probability is 1/6 that a random pairing of the two sets of three scores will provide a perfect positive rank correlation. In pairing problems of this kind there are $n!$ (n factorial) different ways to pair two sets of n scores each.

Example 2.9

Consider a different kind of problem, one concerning combinations. Randomly divide seven scores, all different, into two sets, one set to contain two scores and the other set five scores. The scores will be represented by rank numbers 1, 2, 3, 4, 5, 6, and 7, where rank 1 refers to the largest score. Random selection of two scores (without replacement) implies that each of the seven numbers is equally likely to be drawn on the first draw, and that each of the remaining six is equally likely to be drawn on the second draw, which can be represented in this fashion:

```
1st draw      1          2          3          4          5          6          7
2nd draw  234567    134567    124567    123567    123467    123457    123456
```

There are 42 equally likely events represented, each consisting of two scores. What is the probability that two scores selected at random from the seven different scores will be the two largest? In other words, what is the probability of getting ranks 1 and 2 in the random selection of two scores from the seven? Two of the 42 events consist of 1 and 2: 1-2 and

2-1. Therefore, the probability is 2/42. In combination problems of this kind, there are

$$\frac{n!}{r!(n-r)!}$$

equally likely events, where n is the number of things to be divided and r is the number of things to assign to one of the two sets. It does not matter whether r is the number of things in the larger or the smaller set—the value of the denominator is the same.

Chapter Three

Sampling

In the previous chapter it was noted that an essential step in constructing a probability model for a particular situation is the determination of equal likelihoods in a manner that is generally acceptable. For sampling, this means using a procedure that can be considered a random sampling procedure. It will be shown that the lottery is a random sampling procedure that is an ideal procedure, a criterion by which other sampling procedures can be judged. Common sampling practices will be examined to see how they might be performed by means of a lottery.

Random sampling

Statistical inference based on samples presupposes random sampling. What is random sampling? It is sometimes defined as a procedure of sampling that gives every individual in the population the same chance of being selected. In the preceding chapter, however, it was noted that the "equal chance" of an event occurring is not objectively determined but is another way of saying that the events (in this case, the selection of individuals) are *judged* to be equally likely. It was indicated that the procedure of determining equal likelihoods should be a generally acceptable one in order that computed probabilities will be of significance to other persons. Equal likelihoods for sampling refers to the equal likelihood of each individual being selected, and so a generally acceptable sampling procedure will provide general agreement that one individual is as likely to be selected as another.

Some sampling procedures that are judged by some persons to provide equal likelihood of selection for all individuals are judged by a number of other persons to provide *unequal* likelihoods. Such procedures are, therefore, of little use in making statistical inferences to communicate to others. The concern in this chapter is with sampling procedures that would be generally acceptable; that is, procedures for which there would be general agreement that the notion of equal likelihood of selection of all individuals is justified. A carefully designed lottery is such a procedure. Each individual in the population can be represented by a pellet in a container and a sampling of the individuals can be simulated by taking pellets from the container. Various precautions in the conduct of lotteries can be taken to ensure the reasonableness of the concept of equal likelihood. For instance, one can make sure that the pellets are alike, so that a difference in size or weight or some other attribute of that nature does not cause some pellets to be favored over others. Also, it is necessary to ensure adequate mixing of the pellets. And, of course, it must be ensured that the person drawing the pellet cannot know in advance which individual the pellet represents.

According to some persons, a procedure is a random sampling procedure only if all possible *combinations* of individuals are equally likely. This, the conventional statistical criterion, will be followed in this book. The lottery, of course, meets this requirement.

One of the advantages in using the lottery is that previous uses of lotteries have shown what conditions to control. The use of similar containers, pellets, and procedures over a period of time has allowed improvements in lotteries. Because extensive experience with lotteries has provided empirical evidence of the closeness of the sample distributions to theoretical distributions from probability models and because knowledge that the relevant factors are controlled in lotteries makes it plausible to accept the notion of equal likelihood, the well-conducted lottery would be acceptable to virtually everyone as a random sampling procedure which can be represented by a probability model.

For the reasons just given, we will consider a well-conducted lottery to be a random sampling procedure. Furthermore, whether one uses tables of random numbers, dice, or some other procedure in sampling we will presume that such procedures are never more valid sampling procedures than a lottery procedure, although they may well be more expedient. The lottery, therefore, will be the procedure that is recommended for use in obtaining random samples and in thinking about random sampling. This necessarily restricts the concept of random sampling to finite populations, since there must be a finite number of pellets from which to select. But this restriction has no effect at all on actual sampling because existing populations of discrete elements always contain a finite number of elements and because so-called continuous populations such as areas can be subdivided into a finite population of elements. The concept of an infinite population will be considered to have no relevance for random sampling or statistical inferences based on random sampling.

Measurements are not randomly sampled

Recognition of the finiteness of any randomly sampled population focuses attention on a question that does not arise when one conceives of the population as infinite: Why not take all of the data in the entire population since you have to identify every individual in the population in order to take a sample? If one conceives of taking a random sample of a finite population of *measurements,* there is no good reason for taking a sample rather than using the entire population of measurements. If the pellets in the finite population indicated measurements, one could have simply recorded each measurement at the time he would have coded the pellet and have computed the parameter directly from the recorded measurements. If, then, we actually had records of measurements readily available on all the individuals in the population, we could refer directly to those records instead of taking a sample of individuals from the records. In such a case there would be no reason to take a sample. When we must rely on a sample it is because we do *not,* in fact, take a random sample of *measurements,* but of pellets representing *individuals* who must be measured after being randomly selected.

Reasons for not studying entire populations

Granting the necessity of obtaining a measurement from each individual, why not study the entire finite population and take measurements on each member rather than study just a portion of the population?

Wallis and Roberts (1956) present several reasons for taking a sample instead of studying the entire population. First, there is the expense involved: not only travel and similar expenses, but also the expense of *measurement* (Wallis and Roberts, page 112). Even if it were otherwise possible, the cost would prohibit a study of the yield of every acre of millions of acres of corn or a study of the amount of time every housewife in the United States spends weekly at grocery shopping. A second reason for not studying an entire population is that it may take too much time (Wallis and Roberts, page 112). Extra time means extra dollars, but extra time may be costly in other ways, too. For instance, if we want to poll voters as close as possible to election time the shortness of the time it takes to obtain a sample may be extremely important. A

24 Statistical Inference: The Distribution-free Approach

third reason for sampling instead of studying the complete population is that when measurements are made on every individual in a population many of the measurements may be invalid, making the computed parameter value further from the actual value than a sample estimate would be (Wallis and Roberts, page 113). The reason is that sometimes it is possible to get a few competent persons for obtaining data in a sampling study but impossible to get the large number of competent persons needed to make an accurate complete census. The small number of good measurements may provide more useful information than the much larger number of measurements made by less competent persons. That is the reason the U.S. Census Bureau uses samples to check the accuracy of the census, although sampling cannot replace the census at the present time because the United States Constitution requires a complete census (Wallis and Roberts, page 114). A fourth reason for taking samples is that study of the entire population is inappropriate when the study damages or destroys the individual units being studied (Wallis and Roberts, page 112). Some samples are selected for testing items and when the items to be tested are destroyed by the testing process, as in tests of the breaking point of construction materials or in tests of explosives, testing the entire population is obviously a poor practice.

Analysis of variability: the key to efficient sampling

A more efficient sampling procedure is one that in general with the same sample size provides a more precise estimate of a population parameter. The following discussion of sampling efficiency concerns sampling for the purpose of estimating a population mean only; estimation of other parameters is not treated.

Isolation of sources of variability in a population is the basis of efficient sampling. Principles of sampling design can be understood by realizing how assumptions or knowledge about population variability have been taken into consideration in the design. It should not be surprising that sampling procedures are linked to an analysis of variability; after all, a sampling problem does not exist when a population is known to be absolutely homogeneous, i.e., when a population is known to have no variability.

If a population were known to be absolutely homogeneous with regard to the variable being estimated, one could make just as good an estimate of the population mean from a sample of one individual as from a sample of ten thousand. Increasing the sample size would not provide any new information because one would know beforehand that each of the additional individuals would provide the same measurement. In such a case there would be no sampling problem; one sampling procedure would be as good as another.

Thus, from absolutely homogeneous populations, repeated samples, each consisting of only one individual, will give the same estimate repeatedly. For this reason, increasing the sample size would have no effect on the precision of the estimate. When there is some degree of variability, however, an increase in the sample size tends to increase the precision of the estimate.

Example 3.1

Consider the theoretical sampling distribution of all equally probable samples from the finite population of the first 100 integers, for samples obtained by random sampling without replacement. For samples of one score there would be considerable variation among the sample means, each sample mean being the integer itself. The possible means range from 1 to 100. By contrast, for samples of 99 integers the means would be almost the same. In fact, the smallest possible mean would be 50, obtained when the sample contained the numbers 1 to 99, and the largest possible mean would be 51, obtained when the sample contained the numbers 2 to 100. Sample sizes between 1 and 99 would provide intermediate degrees of variation among the sample means. Now, large variation in the sample mean from one sample to another indicates a low probability that a single randomly selected sample has a mean close to the population mean. Anything that provides greater consistency of means from sample to sample provides a higher probability of a single sample giving a good estimate, an estimate close to the population mean. For all but completely homogeneous populations, larger samples provide greater consistency of estimation from sample to sample and, consequently, better estimation of population means.

Besides indicating the influence of the sample size on the precision of estimation, the preceding example also suggests the influence of variability within a population on precision of estimation. A population with no variability at all gives a perfect estimate of the mean with every sample, whereas for variable populations there is some variability in the sample mean and, consequently, some inaccuracy in the estimate of the mean. The less the variability within a population the more accurate the estimate of the mean.

Random sampling techniques

Statistical inferences about population parameters must be based on random sampling; without random sampling, confidence statements are not justified. But random sampling is the basis of various sampling techniques, and some are more efficient than others.

The term "random sampling technique" will be used to refer to a systematic sampling technique based on the application of random sampling. The easiest part of the technique to perform is the selection of individuals by means of the lottery; a more difficult part of the technique is the planning that precedes the use of the lottery.

The most common random sampling technique is *simple random sampling,* in which a sample size is set in advance and then a sample is taken from a single container of pellets representing the individuals in the population. This technique has the advantage of simplicity, but *sequential sampling* and *stratified sampling* are more efficient techniques. Along with a third random sampling technique, *cluster sampling,* they will be discussed in the following sections. Each of these employs the lottery in a different way from simple random sampling. In sequential sampling the sample size is *not* specified in advance. Stratified sampling requires not one but several containers of pellets, each container representing a *subpopulation* to be sampled. The population itself, however, is not randomly sampled. And in cluster sampling, each pellet within a container represents a specific group or cluster of individuals, not a single individual.

Variability within a population, and sample size, both affect the precision of estimation of a mean, and both play a prominent role in the following discussions of sequential, stratified, and cluster sampling.

Sequential sampling

The more variable a population is, the less precise the estimation of the mean for any particular sample size. It is possible, however, to compensate for greater variability by increasing the sample size, since larger samples provide more precise estimation. If the population variability could be accurately judged in advance, the sample size could be set to provide the desired degree of precision, but the variability would rarely be known without the mean also being known, in which case it would be unnecessary to estimate the mean. A technique known as *sequential sampling* can be used when lack of knowledge of population variability makes it difficult to judge how large a sample is required. In sequential sampling the sample size is not specified in advance. Prior to selecting the sample, a criterion for the precision of estimation is established, and then sampling is continued until the criterion is met. For instance, suppose that in an estimation of the mean weight of a population of animals the criterion is a 95 percent confidence interval with confidence limits no more than 5 pounds from the sample mean. Sampling is continued until the confidence interval is that small. In general, the more variable a population the longer the sampling continues. Sequential sampling can be economical when a population has small variability because the sample size required to meet the criterion may be considerably smaller than the sample size that would have been set in advance. It should be noted, however, that sequential sampling may exhaust a finite population. This consideration does not invalidate sequential sampling, however, because the criterion for precision of estimation is certain to be met when the population is exhausted.

Stratified sampling

Stratified sampling is a technique based on separate random sampling of subpopulations (called strata) of the population whose parameter is to be estimated. An example will show the basic difference between the simple random sampling model and the stratified sampling model.

The total variability of a population can be considered to consist of two parts: the variability within the individual strata and the variability between them resulting from the difference between their means. When there is a difference between the stratum means that is large compared to the variation within strata, by sampling the strata separately to obtain estimates of stratum means and combining these estimates one should get better estimates of population means than by simple random sampling of the total population. This is because separate sampling of the individual strata prevents the difference between the stratum means from contributing to the variability of the estimates. An example will illustrate this point.

Example 3.2

Consider the model for a simple random sample of two objects taken without replacement from a population of five objects, A, B, C, D, and E. The object of the sampling is to estimate the mean weight of the five objects. The weights of the objects (unknown to the investigator, of course, at the time of sampling), in pounds, are A: 10; B: 20; C: 70; D: 80; and E: 90. Random sampling of two objects without replacement

28 Statistical Inference: The Distribution-free Approach

from this population can be represented by this diagram which shows the weights of the equally likely pairs of objects and the means of the equally likely pairs:

1st draw	10	20	70	80	90
2nd draw	20 70 80 90	10 70 80 90	10 20 80 90	10 20 70 90	10 20 70 80
Sample means:	15 40 45 50	15 45 50 55	40 45 75 80	45 50 75 85	50 55 80 85

Notice that the sample means vary from 15 to 85 pounds. Since the sample means are the estimates of the population mean, this indicates the variation from sample to sample in estimates of the population mean.

Example 3.3

Stratified sampling of the same population will now be represented in a diagram. Even without weighing the objects, the investigator suspects that objects A and B are considerably lighter than objects C, D, and E. He therefore decides to obtain two separate estimates, an estimate of the mean of the subpopulation consisting of objects A and B and an estimate of the mean of the other subpopulation consisting of objects C, D, and E. He randomly selects one object from the A-B subpopulation and one object from the C-D-E subpopulation. The following diagram represents the equally likely pairs of objects obtained by this procedure:

A-B object	10	20
C-D-E object	70 80 90	70 80 90
Population mean estimates:	46 52 58	50 56 62

The population mean estimates are obtained by combining the weights of the sample objects in this manner: Multiply the weight of the A-B object in the sample by 2 and add the result to the weight of the C-D-E object multiplied by 3 to get an estimate of the total weight of the five objects; then divide this estimate by 5. For example:

$$\frac{(10 \cdot 2) + (70 \cdot 3)}{5} = 46$$

Although the estimates of the population mean with simple random sampling in Example 3.2 varied from 15 to 85 pounds over the various equally likely samples, the sample-to-sample variation for stratified sampling is considerably less, the smallest estimate being 46 pounds and the largest 62 pounds. This example does not *prove* that stratified sampling is always more efficient than simple random sampling, but it does show something of the reasoning behind stratified sampling. Stratified sampling is most effective when there is little variability within strata and much variability between strata. By separate sampling of quite different strata one reduces the variability among population mean estimates by eliminating the effect of between-strata variability; the variation among samples is then entirely the result of within-strata variability.

Example 3.4 will clarify other aspects of stratified sampling.

Example 3.4

Johnson has two plots of land, plot *A* and plot *B*. He wants to estimate the mean number of bushels of grain per acre. On a map he divides each plot into one-acre units to allow taking a random sample of the units. Suppose that the plot *A* acres have yields between 50 and 60 bushels per acre and that the plot *B* acres have yields between 100 and 120 bushels per acre. (Johnson at the time of the sampling does not know these yields, of course, but he knows that plot *B* has much better soil, and so he does expect it to have a considerably higher yield.) The variability within plot *A* and within plot *B* is small but the variability within the population as a whole (both plots combined) could be considerable because of the large difference in yield between the two strata. A simple random sample from the entire population probably would give a mean that differed greatly from the mean of another simple random sample because a change in the proportion of plot *A* or plot *B* acres from one sample to the next could shift the sample mean (which is the estimate of the population mean) up or down by a great amount. On the other hand, sampling each stratum separately would give essentially the same mean for repeated sampling within each of the strata. Consequently, the estimates of the mean of plot *A* and the mean of plot *B* would usually be quite precise. How can these two estimates be combined to get an estimate of the *population* mean? The first step is to weight each sample mean by the number of acres in the stratum, which provides two estimates of *stratum* totals. Then the two weighted sample means are added to give an estimate of the population total. This estimate is then divided by the number of acres in the population, which is the total number of acres in the two strata, to get an estimate of the population mean. For instance, if the population consisted of 50 acres in plot A and 100 acres in plot B and if the plot A sample mean was 54 and the plot B sample mean was 112, the estimate of the population mean would be

$$\frac{(50 \cdot 54) + (100 \cdot 112)}{150}$$

which is 92 2/3 bushels per acre.

In order to make the argument easier to follow, the preceding example hypothesized distinct strata in which the measurements in the two strata did not overlap at all. It must be emphasized, however, that the technique depends in no way on the assumption that the stratum measurements do not overlap in size. For stratified sampling to be useful it is adequate that the difference between the stratum means be large relative to the variability within strata.

The appropriate way in stratified sampling to combine the stratum sample means in order to estimate the mean of the total population has been discussed, and so the next topic to treat is the other aspect of stratified sampling, the determination of the relative sample sizes for the strata. The stratum sample means are estimates of the stratum means. The error in estimating the population mean depends on the errors in estimating the means of both strata. The appropriate ratio of sample sizes therefore is determined by considering the relative importance of error in the two stratum mean estimates in combining them to get an estimate of the population mean.

Example 3.5

A population of individuals is sampled by randomly sampling stratum 1 and stratum 2 separately. Stratum 1 contains 1,000 individuals, whereas stratum 2 contains only 100. The estimate of the population mean will be

$$\frac{1{,}000\ \bar{X}_1 + 100\ \bar{X}_2}{1{,}100}$$

where $\bar{X}_1$ and $\bar{X}_2$ are the stratum sample means, which are the estimates of the means of the corresponding strata. The variation in $\bar{X}_1$, the estimate weighted by 1,000, has more effect on the variation of the population mean estimate than does variation in $\bar{X}_2$, the estimate weighted by only 100. Consequently, more is accomplished by minimizing the error in the stratum 1 mean estimate than by minimizing the error in the mean estimate of the smaller stratum, stratum 2. A measure of the error associated with a mean estimate is the *standard error of the mean,* which is directly related to the population standard deviation and inversely related to the sample size. In the standard error of the mean, therefore, we have an expression that combines the two factors whose effects on the accuracy of estimation were discussed earlier: variability within a population and sample size. For the standard error of the mean of a *stratum,* of course, it is the *stratum* variability which is the "population" variability involved. The stratum variability is fixed, but the standard error of the mean of the heavily weighted stratum can be reduced by increasing the stratum sample size, less increase being needed when the stratum variability is small than when it is large. It has been algebraically demonstrated by mathematicians that the estimate of a population mean based on stratified sampling has the greatest precision (smallest standard error) when the sample sizes for the strata are in the same ratio as the products of the standard deviation of the stratum and the stratum size.[1]

Example 3.6

For example, suppose that stratum 1, with 1,000 individuals, has a standard deviation of 5 and stratum 2, with 100 individuals, has a standard deviation of 20. The ratio of the products of the stratum sizes and standard deviations is (1,000 · 5):(100 · 20), or 5:2. The optimum ratio of sample sizes therefore is 5:2. In a sample of 70 individuals, for example, 50 should be taken at random from stratum 1 and 20 at random from stratum 2. When, as in this case, the ratio of sample sizes is *not* the ratio of the stratum sizes, the sampling procedure is known as *nonproportional stratified sampling.*

---

[1]Actually, the sample sizes for the strata should be in the same ratio as the products of the standard deviation of the stratum and

$$\sqrt{\frac{N^3}{N-1}}$$

where N is the stratum size (Kendall, 1943). Using N as an approximation to

$$\sqrt{\frac{N^3}{N-1}}$$

reduces the efficiency only slightly and does not bias the estimate at all.

Stratified sampling is more efficient when the stratum variability is known but, as was pointed out in the discussion of sequential sampling, it is unlikely that a person would know the variability of a population or stratum without also knowing the mean, in which case there would be no estimation problem. Nevertheless, the rule for determination of the optimum sample size can be useful even when the stratum variability is unknown, because it may be possible to *estimate* the relative sizes of the standard deviations of the strata and adjust the sample sizes accordingly. Or when there is no basis for estimating their relative variability, the strata can be treated as equally variable. In this case the ratio of the products of the stratum size and the standard deviation of the stratum is simply the ratio of the stratum sizes.

Example 3.7

In the preceding example if both stratum standard deviations had been the same, say 25, the optimum ratio of the sample sizes would have been (1,000 · 25):(100 · 25), or 1,000:100. In a sample of 110 individuals, for example, 100 should be taken from stratum 1 and 10 from stratum 2. When the strata are treated as equally variable, either because it seems plausible that they should be equally variable or because of ignorance of the relative variability, the optimum ratio of the sample sizes is the ratio of the stratum sizes. Notice that there is *no bias* introduced into the estimate of the population mean by treating the standard deviations of the strata as equal even when they are quite different; weighting the stratum mean estimates by the stratum size before adding them guarantees lack of bias in the estimate of the population mean. When the ratio of sample sizes is the ratio of the stratum sizes the stratified sampling procedure is known as *proportional stratified sampling*.

A type of *nonproportional* stratified sampling which does not take into consideration the relative stratum variability is sometimes used, although it has no logical justification. This procedure involves taking larger samples from strata which have the larger measurements when the object is to estimate the total of the measurements, such as the total monetary value of an inventory.

Example 3.8

For example, in taking inventory, it might be proposed that larger samples should be taken from the stratum of $40–$50 items than from the stratum of $30–$40 items because the higher-priced items contribute more to the total. This proposal is not logical because the important consideration for estimation is not how much a particular item's value contributes to the total value but how much the estimate of the total value is improved by knowledge of that item's exact value. A sample from the $40–$50 range improves estimates of the total value of the inventory no more than a sample of the same size from the $30–$40 range if the strata are the same in size and variability. For instance, if it were known that there was only one item in each category, the total value could not be estimated any better from knowledge of the exact value of the item in the higher price range than from knowledge of the exact value of the item in the lower price range. Similarly, if there are 20 items in each of the two price ranges, knowledge of the total value of all 40 items is increased no more by taking a sample of a given size from the $40–$50 range than by taking a sample of the same size from the $30–$40 range. Thus, the sample size for a stratum is independent of the value of the items in that stratum. It frequently happens, however, that one *should* take large samples from the strata with the higher values. This is because it is common to form strata with wider ranges for higher-priced items, for example, $5–$10, $10–$25, $25–$50, etc. In such cases samples larger than proportional samples should be taken from the strata with the higher values but only because those strata have wider ranges of values and therefore are more variable.

Cluster sampling

When the individuals in a population form convenient groups or clusters, sampling may be carried out by taking a random sample of clusters and obtaining measurements from the individual elements in the clusters. This is known as *cluster sampling*.

Example 3.9

Mr. Cole, a city planner, wants to estimate the mean income of the families in his city. He takes a random sample of 50 blocks and interviews all of the families in each of the sample blocks to determine the family income. The mean number of families per block in the sample blocks is 10, and so altogether 500 families were interviewed. He divides the total income of the 500 families by 50 to get the mean income per block. Then he converts the mean income per block into an estimated total income for the city by multiplying it by the number of blocks in the city. The estimated total income is then divided by the total number of families in the city to get an unbiased estimate of the mean income per family.

Notice that Mr. Cole obtained an estimate of the mean income of families in the city without taking a random sample of the families. The probability model that can be used to show the relationship between random sampling of *blocks* and inferences about *families* should be discussed. The model is a random sampling model where there is a population of *blocks* and each block is equally likely to be selected. The total income in a block can be regarded as a measurable characteristic of the individual block in the same sense that individual income is a measurable characteristic of an individual person. Since the blocks are equally likely to be selected, the mean income of the sample blocks is an unbiased estimate of the mean income of all the blocks in the population. The mean income of the sample blocks is simply the total income of the sample blocks divided by the number of blocks in the sample. Multiplying this unbiased estimate of the mean income of all city blocks by the number of blocks in the city provides an unbiased estimate of the total income of all city blocks, which is the total income of all of the families in the city, since every family resides in a city block. From the unbiased estimate of the total income of all families we can get an unbiased estimate of the mean income per family by dividing by the number of families in the city.

Now it might seem that the procedure used by Mr. Cole in estimating is unnecessarily complicated, that he could simply have used the mean income of the families over all the blocks in the sample as an unbiased estimate of the population mean family income. In support of this view is the fact that although blocks, not families, were randomly selected, Cole ensured equal likelihood of selection for every family in the city by interviewing all of the families in each sample block. All families were equally likely to be selected because every family resides in a block that has the same probability of being selected as any other block and because selection of a block is selection of all of the families in the block, regardless of the number of families. In spite of this fact, the mean income of families over the blocks in the sample is *not* an unbiased estimate of the mean income of all families in the city. The following example will show why.

Example 3.10

Suppose there were only two blocks, block A and block B, in the city. One of the two blocks is randomly selected in order to estimate the mean family income over both blocks. Block A has 40 families with a mean income of $2,000 whereas block B has a mean income of $20,000 and contains only 10 families. The two equally likely samples can be represented in this way:

Block A	Block B
Mean = $ 2,000	Mean = $ 20,000
40 families	10 families

If the investigator uses the mean family income in his sample as an estimate of the mean family income in the population, the probability is 1/2 that he will estimate $2,000 and 1/2 that he will estimate $20,000. The mean of these equally probable estimates is $11,000, whereas the actual population mean family income is

$$\frac{40(\$2,000) + 10(\$20,000)}{40 + 10}$$

—that is, $5,600. Thus, the sampling technique is a biased one with a tendency toward overestimation. When block *A* is selected the population mean is *underestimated* by $3,600, whereas when block *B* is selected the population mean is *overestimated* by a much larger amount, $14,400. The source of the bias is this: Since the number of families in a block is not taken into consideration in the estimate the mean from block *A*, based on 40 families, is given no more weight than the mean from block *B*, based on 10 families.

You can get an intuitive grasp of the nature of the bias if you imagine yourself in a situation in which every block in your city gets one vote regarding a bill on family allowances. You live in a crowded block and therefore do not think it fair that your block is represented by a single vote, the same as a block with very few families. Then logically you should also consider unfair an alternative procedure in which a random sample is taken of all the blocks in the city and the blocks in the sample are each given a single vote. What is unfair in the case of the largest possible sample, namely, the population, is also unfair for smaller samples.

When simple random sampling is used, all combinations of individuals are equally likely to be drawn. Cluster sampling is a simple random sampling technique in terms of clusters, but not in terms of the individuals within the clusters. The example of bias just considered illustrates that equal likelihood of individuals being drawn does not imply equal likelihood of all *combinations* of individuals in cluster sampling. In the preceding example, only two combinations of the 50 families could possibly be selected.

There is a sampling and estimation technique based on clusters that is probably more commonly used than the technique recommended in Example 3.9. It involves taking a random sample of clusters and then, instead of obtaining a measurement from each individual in the selected clusters, measuring a random sample of individuals from each of the randomly selected clusters, taking the same size sample from each cluster. Then all of these measurements are averaged over all of the sample clusters to get an estimate of the population mean. This is a biased procedure. Whereas the mean of a random sample of elements from a larger random sample of *elements* is an unbiased estimate of the mean of the population of elements, the mean of a random sample of elements from a random sample of *clusters* is a biased estimate of the mean of the population of elements. The following example will illustrate the bias.

Example 3.11

We will use the population of two blocks considered in Example 3.10 with the restriction that all of the families within a block have *identical* incomes. A random sample of one block is taken and then a random sample of five families is taken from that block. The mean income of the five families is used as an estimate of the mean family income over both blocks. Because of the complete homogeneity of family incomes within blocks, there will be two equally probable estimates, just as in Example 3.10, an estimate of $2,000 every time block *A* is selected and an estimate of $20,000 every time block *B* is selected. Therefore, the bias is exactly the same as in Example 3.10, the mean of all equally probable estimates being $11,000 in contrast to the actual mean of $5,600.

Example 3.12

Another way of looking at the sampling procedure used in Example 3.11 will show that the bias exists even when the families within blocks do not have identical incomes. No matter what the degree of variability of incomes within blocks, the sampling procedure provides equally probable estimates from block *A* which average out to $2,000 and equally probable estimates from block *B* which average out to $20,000. Since a block *A* estimate and one from block *B* are equally probable, the mean of all equally probable estimates from both block *A* and block *B* is $11,000.

The 500 families interviewed by Mr. Cole would not provide as accurate an estimate of the population mean family income as would 500 families drawn at random from the entire city. As Wallis and Roberts (pages 487-488) have pointed out, families in the same block are likely to have very similar incomes, and so after the first family in a block is interviewed the information from the other families to be interviewed does not add much to the information obtained from the first family.

Simple random sampling is more efficient than cluster sampling in terms of the precision of estimate for a given sample size. But cluster sampling may be so much easier to perform that it costs less to get a good estimate by means of a large cluster sample than to collect the smaller random sample required for the same precision of estimation.

Clusters are formed on some basis that reduces cost, often geographical grouping (Wallis and Roberts, pages 487-488). Geographical proximity frequently implies considerable homogeneity within clusters, as in the above example. In contrast to stratified sampling, cluster sampling is most useful when there is *much* variability within clusters and *little* variability between clusters, because a few clusters with many measurements would then provide a good estimate of a population mean.

Stratified cluster sampling is, of course, possible. For instance, in estimating the mean family income in a city, more efficient estimation can be gained by dividing the city into two or more strata which are expected to have quite different mean family incomes. Each stratum is randomly sampled by the cluster sampling procedure, setting the sample size (number of blocks) for a stratum according to the proportion of the city blocks in that stratum and the relative variability of block means in that stratum (if knowledge of this variability is available) as specified in the preceding section on stratified sampling. For each stratum an estimate of the mean block income for that stratum is obtained. This estimate is multiplied by the number of blocks in the stratum to provide an estimate of the total income in that stratum. The estimates of total stratum income are added to provide an estimate of the total income over all strata, which is, of course, an estimate of the total family income in the city. Dividing this estimate by the number of families in the city provides an unbiased estimate of the mean family income in the city.

Sources of variability

Knowledge of sources of variability within a population has just been shown to allow improvement in sampling designs. That is one reason an expert in a field can sample more efficiently than other persons. The economist knows how to stratify a population to provide a good estimate of personal income; the ornithologist knows how to stratify a geographical area for estimating the number of birds; and the agriculturalist knows how to stratify for estimating wheat production. A person who is not an expert in a field can, however, frequently improve the efficiency of a sampling design by considering carefully the likely sources of variability among the individual elements in the population.

One of the sources of variability among individuals within a population is the difference in their origins. Variation in genetic structure among plants and animals can account for heterogeneity of size, shape, color, etc. For inanimate objects the counterpart of genetic variation is the process of formation of the objects and variation in the materials composing the objects. For example, rocks differ in hardness because some are formed by sedimentation while others are formed by fusion of the materials by heat, and also because of differences in the materials composing the rocks.

Another factor that produces variability is the variation in the environments in which the individuals are located. For all sorts of animate and inanimate things, there are differences which result from variation in the physical environment. Genetically identical plants in dry and wet soil may differ in size and appearance. One rock is less exposed than another to rain and sunshine and consequently differs in appearance and hardness as a result of the difference in exposure. In humans, not only do physical size and appearance depend on the physical environment, but so also do personality characteristics. The symbolic environment as well as the physical environment must be taken into consideration as a factor producing differences among humans. Differences in the cultural and social environments, regardless of the homogeneity of the physical environment, cause heterogeneity of attitudes, intelligence, and other significant attributes of humans.

By carefully considering the possible causes of variability of the characteristic being estimated, even a nonspecialist may be able to stratify the population in a way that improves the efficiency of estimation.

Indirect sampling

Indirect sampling can sometimes lead to bias. In indirect sampling a researcher takes samples from a population different from the one in which he is primarily interested. The following example, based on an example from Wallis and Roberts (1956, page 72), shows how indirect sampling can lead to biased estimation.

Example 3.13

In order to estimate the average age of fathers of schoolchildren, a researcher may question a random sample of schoolchildren instead of a random sample of the fathers. The estimate obtained from the mean of the ages given by the children is probably an overestimate because there is a greater probability of selecting a child from a family with several schoolchildren than of selecting a child from a family with fewer schoolchildren and because the fathers in a family with several schoolchildren are likely to be older than the fathers of fewer schoolchildren. Consider a numerical example. For only two families, one with six schoolchildren with a father aged 40, and one with one schoolchild with a father aged 26, the average ages of the fathers is 33, but if each of the seven children were asked the age of his father, the average of the seven replies would be 38. Now it may appear that one could avoid this bias by simply not taking in his sample more than one child from a family, but this is not a solution because there still would be a greater probability of selecting a child from a large family than a child from a small family and, consequently, a greater probability of selecting a child with an older father than one with a young father. However, an adjustment can be made on data obtained in this fashion to compensate for the bias. In order to compensate for the bias, it is necessary to find out from each child questioned, the number of schoolchildren in his family as well as his father's age. Suppose the following data were obtained from questioning 50 schoolchildren:

Size of family of respondent	Number of respondents	Average age of father
One-schoolchild family	11	24
Two-schoolchild family	24	26
Three-schoolchild family	15	30

Without adjustment, the average of the ages is 11(24) + 24(26) + 15(30), divided by 50, which is 26 38/50. An estimate obtained by such a procedure is likely to overestimate the population mean. The fathers of the larger families are older and are more likely to contribute to the sample because they have more children that can be selected. More specifically, the probability of a particular father of one schoolchild contributing to the estimate is one-half that of a particular father of two schoolchildren and one-third that of a particular father of three schoolchildren. A compensation for this bias is to reduce the weight given the average age of the fathers of the larger families to counteract the effect of disproportionate representation of such families in the sample (Wallis and Roberts, page 120). The weight given the mean for two-schoolchild families is halved and the weight given the mean for three-schoolchild families is divided by 3. That is, instead of weighting the two-schoolchild family mean by 24, it is weighted by 12, and instead of weighting the three-schoolchild family mean by 15, it is weighted by 5. Thus, the *adjusted estimate* is 11(24) + 12(26) + 5(30), divided by 28, which is 25 26/28.

Wallis and Roberts (page 72) pointed out that a similar problem has been encountered at times in estimating family earnings by sampling wage earners listed in the employment records. Families with more than one wage earner have a higher probability of being included in the sample. Multiple-earner families tend to have higher incomes than single-earner families because of the multiplicity of earners. An adjustment analogous to that used in the fathers' age example could be made for family earnings by giving weight to earnings inversely proportional to the number of wage earners in the family, when the number of wage earners in each family could be readily determined.

Sampling subpopulations

Sampling a subpopulation or selecting all individuals in a subpopulation is sometimes done in order to make "statistical" inferences about the entire population. In some cases it is not possible, or at least not practical, to sample the entire population whose parameters are to be estimated, although a subpopulation can be sampled. The estimate for the subpopulation is sometimes treated as an unbiased estimate for the entire population of which the subpopulation is a part. The procedure of using a random sample from a subpopulation to make inferences about a population parameter is frequently a biased procedure.

An example of bias that can result from sampling a nonrepresentative subpopulation is the estimation of certain measurements when there has been inadequate sampling over a period of time. For example, to estimate the average amount of time workers require for a particular task, it is desirable to study the performance at random intervals during a span of time. Time-study specialists have found that work production during certain times of the day or certain days of the week deviates considerably from the average daily or weekly production, and so bias can easily occur if observations are not drawn from the entire span of time. Employment also varies over the week. Unemployment figures have occasionally been based on counts of workers made on a single day of the week, in some cases Monday (Wallis and Roberts, page 70). But frequently part-time workers work on the early days of the week but not on the later days, and so an estimate based on the number of workers on a Monday might be an overestimate of the average number of workers on a workday.

Another source of bias in connection with sampling a time span concerns the error in treating early returns from respondents as typical returns. To illustrate, after the 1930 census it was stated on the basis of returns from one-half of the states that the 1930 divorce rate had fallen from the rate in 1920 (Wallis and Roberts, page 76). However, when results from the rest of the states were tabulated it was found that the rate had not changed from the 1920 rate. The divorce rate for 1930 was underestimated because the first states reporting were the less populated, agricultural states, which usually have lower divorce rates. Having fewer records to contend with, these states could tabulate their data quicker.

In some cases bias results from the difficulty in locating certain individuals in the population. The difficulty of locating some individuals can be so great that they can be considered virtually unobtainable. For instance, we may not be able to take a random sample of trees in a jungle because some parts of the jungle are, for all practical purposes, impenetrable. Samples of minerals from steep slopes of mountains may also be practically impossible to obtain. Similarly, there are places on the ocean bottom that are so difficult to sample because of depth or turbulence that they would be avoided in a sampling study. For various reasons it is sometimes difficult to locate certain types of people. A census taken for military and taxation purposes may show a population that is much smaller than is shown by a census taken for the purpose of providing federal aid because people are easier to count when they want to be counted. Even where there is no effort by the individuals to seek or avoid detection there may be a considerable difference in the ease of locating certain types of individuals, which can lead to bias. For instance, there seem to be more cases of mental and nervous diseases among men than among women, but this impression may be incorrect (Wallis and Roberts, page 72). Men are more likely to be detected and institutionalized because such disorders are more likely to interfere with their regular work and because they are not likely to be supported by another member of the family if they are unable to support themselves. This would lead to an overestimation of the male/female ratio for mental and nervous diseases. Individuals in a population are sometimes difficult to locate and are therefore unlikely to be selected in a sample because they are not listed in the records used for sampling purposes. It is a common practice to take samples from records in order to conduct a survey. If a survey contains topics that are likely to receive a different response from people with different incomes, certain records are likely to lead to biased sampling. City directories do not list transients, who frequently are poor. Telephone directories do not show the names of persons who cannot afford a telephone. Real estate tax records do not include the names of permanent residents who rent houses or apartments. The records of a college alumni association can provide the present addresses of a large number of alumni, but the addresses of successful alumni are more likely to be known than those of the less successful.

In all of the preceding examples *random sampling* of a subpopulation would allow statistical inferences about the *subpopulation* parameters, but inferences about the general *population* parameters would require evidence of the representativeness of the subpopulation sampled. Inferences about the population parameters, therefore, would be *nonstatistical inferences,* inferences with no basis in probability theory. In contrast to statistical inferences about a randomly sampled subpopulation these nonstatistical inferences about population parameters may not appear to other people to be justifiable inferences. A person who has randomly sampled a subpopulation and wants to draw inferences about a population should present his arguments for generalizing from the subpopulation to the population.

Sampling to estimate the size of animal populations

Random sampling for the purpose of estimating the size of an animal population requires a little thought. Obviously, when the size of the population is unknown it is impossible to represent each animal with a pellet in a container in order to permit random sampling. Instead, an area is subdivided into units, which are represented by pellets to be randomly drawn from a container. A random sample of the population of units is taken and the number of animals in each unit is treated statistically as a property of the unit. If we have taken a random sample of 1/20 of the units, an unbiased estimate of the total number of animals in all of the units is obtained by taking the total number of animals in our sample and multiplying this by 20. We have then an unbiased estimate of the total number of animals in the population contained in the area which was subdivided into units.

Since the animals may not be uniformly dispersed over the area to be sampled, it is better to use a large number of small units of area than a small number of large units. The greater the heterogeneity of sample unit frequencies among the units, the larger the number of sample units required for the same degree of accuracy of estimation of population size.

Stratified sampling is, of course, possible, too (Andrewartha, 1961, page 24). The population of area units is subdivided into subpopulations likely to differ considerably in the mean number of animals per unit.

Taking a census of the animals in the sample units underestimates the population size if the animals are difficult to detect. Salt (1948) as cited in Andrewartha (page 23) estimated the number of arthropods in an acre of soil by cutting out cylinders of soil that were 4 inches in diameter and 12 inches deep. He had to invent a special technique for pulverizing the soil without destroying the animals in it. By this method he hoped to obtain practically a full count of the animals. In sifting out the soil, however, some of the animals may have passed through the sieve. (There is a similar bias against catching fish smaller than the mesh of the net). Another investigator, Milne (1943), as cited in Andrewartha (page 23), collected sheep ticks on a woolen blanket as it was dragged over the grass. Even though he dragged the same area repeatedly, he probably did not get all of the ticks and therefore underestimated the population.

Another procedure for estimating the size of an animal population is the *capture-recapture procedure* (Andrewartha, page 24). In the capture-recapture procedure, animals are captured, marked with identification tags, and released. After a period of time more animals are captured and the proportion of these animals that have identification tags is noted. On the assumption that a tagged animal, i.e., an animal that has been caught previously, has the same probability of being captured as an animal that has not been captured before, the proportion of the captured animals that have identification tags is an unbiased estimate of the proportion of the entire population that have identification tags. Since the number of animals in the entire population with identification tags is known, it is possible to estimate the total population of animals. It would, of course, be possible to use a *release-capture procedure* in which tagged animals that are not originally from the population to be estimated are released. However, the released animals, being unfamiliar to the other animals, might not mix uniformly with them. Also, having a different hereditary and environmental background, they might differ from the rest of the animals in their ability to avoid capture. Even when the tagged animals come from the population being sampled, there may be good reason to question the assumption that the probability of recapturing a tagged animal is the same as the probability of capturing one of the other animals. In one test of this assumption a number of traps were placed in a building containing mice (Evans, 1949, as cited in Andrewartha, page 25). Certain mice were captured many times, whereas others were not caught a single time. The difference in capture rates among the mice was too great to be attributed to chance; it was clear that some mice were easier to capture than others. A mouse that was easy to catch would be more likely to be tagged and more likely than another mouse to be captured at a later date. This would tend to cause an overestimation of the ratio of tagged to untagged animals and a consequent underestimation of the population. Another reason that tagged animals might be easier to capture, if the traps were not moved, is that the place of original capture could be a favorite place to which the tagged animals are likely to return. On the other hand, the tagged animals could be harder to capture than the other animals if they tended to avoid the place where they were captured or to avoid the trap wherever it is placed. The possibility of these opposing tendencies does not imply, of course, that they cancel out. A procedure providing some control over these sources of bias is to set several types of traps at randomly selected places in the territory to capture the animals to be tagged, and then to move the traps to other randomly selected sites and release the tagged animals at randomly selected places within the territory. The major problem in capture-recapture sampling is the definition of the population being sampled. If we have a population on a small island where all of the animals can wander around the island and where there are no animals coming to or leaving the island the population is well defined, but when there are no such bounds to limit the movement of the animals, the problem of definition is great. The capture-recapture procedure well illustrates the amount of ambiguity that may be associated with a parameter estimate based on a *nonrandom* sampling procedure.

A method similar in principle to the capture-recapture procedure can be used in estimating amounts of liquids that cannot be directly measured, such as the amount of blood in a living person or the amount of oil or water in an underground pool. The concentration of some soluble substance in the liquid is determined, and then a measured quantity of that substance is added to the liquid. After sufficient time for the substance to spread uniformly throughout the liquid, but before there is time for it to go out of solution, samples are taken to determine the new concentration. The change in concentration resulting from the introduction of the substance allows an estimation of the total amount of liquid, under the assumption that the concentration is uniform throughout the body of liquid. Provided the substance will readily disperse uniformly in the liquid and remain in suspension until the new level of concentration is determined, the estimate should be a good one. Nonetheless, this would be nonstatistical estimation, because no finite population of units would have been randomly sampled. In fact, if the assumption of uniform dispersion of the introduced substance is correct, there is no need to worry about random sampling, because the population is completely homogeneous.

Area sampling

Sometimes, to simplify sampling, an area is sampled to get parameter estimates for the individuals living in the area. Inappropriate use of area sampling for this purpose can lead to biased estimation. An example of an inappropriate procedure follows. Suppose we want a random sample of farmers. To sample the farming area it is divided into units by means of map grids. Each grid line intersection is numbered and numbered pellets corresponding to those intersections are put in a container. A random sample of grid line intersections is then selected from the container. The owners of the land at the sample intersections constitute the sample. This procedure is biased because all ranchers and farmers in the total area are not equally likely to be selected. Farms that are large are more likely to be selected, and so the selected farmers are nontypical, being wealthier, having different farming practices, and differing systematically in other ways from the typical farmer. This is another instance of indirect sampling and shows the same kind of bias as was shown in Example 3.13 for an estimate of the mean age of fathers of schoolchildren.

Bias from sample dropouts

At the beginning of this chapter it was pointed out that it is desirable to make a distinction between random sampling of *individuals* and random sampling of *measurements* and to recognize that in fact we always randomly sample individuals, not measurements. When it is possible that some of the sample individuals will not provide usable measurements, this distinction is especially important to keep in mind. The advantages of random sampling may be virtually nullified by biased elimination of individuals from the sample due to the inability to get valid measurements from them.

Bias from sample dropouts is especially likely to occur in obtaining responses from persons in interviews or questionnaires. The person conducting the survey may be able to obtain a random sample of the population of individuals in which he is interested but not be able to get usable responses from all of the individuals in his sample. There are two principal ways in which this can occur.

First, some persons in the sample may not provide any responses. Occasionally people refuse to be interviewed. And certainly many people do not return questionnaires that have been mailed to them. This may be an important factor to consider in evaluating questionnaire returns because there is sometimes reason to believe that persons not returning questionnaires would have answered the questions in a different manner from those who returned them. For instance, persons with strong political views may be more inclined to return political questionnaires than persons with more moderate views.

The other main reason that an investigator may not be able to get usable responses to his interview or questionnaire from all persons in his random sample is that sometimes the responses are inadequate. For example, some answers are unusable because they are not honest answers but are given in order to please the interviewer, or perhaps to shock the interviewer, as in interviews about sexual behavior. Also, some persons in the sample may not be able to give usable responses even when they strive to give honest answers, because they do not understand the questions or because they do not know how to give appropriate replies. When we discard the responses of such persons, we introduce the possibility of a strong bias because the ability to properly complete a questionnaire or an interview may be correlated with the attribute that is being investigated.

When only a portion of a random sample provides usable measurements, this portion can be regarded as a random sample from the subpopulation of individuals who, if selected in a sample, would give usable responses. That is, such a sample has the same general statistical properties as a sample selected directly from a group containing only individuals who would give usable responses. For example, suppose we had a population of 100 individuals, 70 of whom would give usable responses. We decide in advance to take individuals at random from the population until we get 20 individuals who provide usable responses. The procedure of discarding individuals who do not provide usable responses until we get 20 who do provide usable responses is statistically equivalent to the procedure of selecting 20 individuals at random directly from the subpopulation of 70 individuals who would give usable responses. The theoretical sampling distributions of samples are identical.

We can, then, regard the portion of a random sample that provides usable responses as a random sample from the subpopulation of individuals who would give usable responses. In order to generalize from the estimates for the subpopulation to the general population of individuals originally sampled, there must be logical reasons for considering the subpopulation to be representative of the total population in regard to the responses. This is equivalent to requiring reasons for assuming that the individuals who provided usable responses gave responses typical of those that would have been given by the other individuals if we had been able to get valid responses from them. For example, an investigator should be expected to justify his assumption that persons who did not return his questionnaire would have answered it in about the same way as those who returned it.

Haphazard sampling

Frequently inferences are drawn about populations on the basis of knowledge of individuals whose selection hardly merits the term "sampling." Nonetheless, such inferences are quite common, and so they deserve a comment. This type of inference is prevalent around the time of elections. Judging political views of the nation on the basis of views of one's neighbors can be risky inasmuch as neighbors tend to have similar education, income, attitudes and other characteristics that are conducive to congenial neighborly living; consequently, neighbors have very similar political views, not a cross section of views. A similar type of bias results when physicians, lawyers, psychologists, or other professional people who listen daily to the personal problems of clients draw inferences about the personal problems of the general public. A marriage counselor who generalizes from his experience with persons who deliberately seek professional help with marital problems is likely to underestimate the self-reliance of married people in general. In the same fashion, pediatricians may be prone to overestimate the prevalence of behavioral disorders in children because of their overexposure to such cases.

Haphazard sampling allows no statistical inference at all and depends entirely on arguments with a nonstatistical basis for generalizing beyond the sample.

Chapter Four

Distribution-free Confidence Intervals

In the previous two chapters, arguments were presented for regarding the lottery as the ideal random sampling procedure. In this chapter we will develop distribution-free confidence intervals on the basis of lottery sampling. The populations considered, therefore, are finite and discontinuous.

The procedures in this chapter are not intended to replace existing procedures for constructing confidence intervals but rather to introduce the reader to alternative ways of thinking about confidence intervals. The examples will show both the value and the method of the distribution-free approach to confidence intervals. A range of examples are given to facilitate extrapolation by the reader in extending and modifying these procedures.

The following procedures depend as little as possible on dubious assumptions. A continuous distribution is not assumed because the series of possible measurements from any measuring device form a discrete, not a continuous, series. An infinite population is not assumed because in application confidence intervals concern populations consisting of a finite number of elements. And the shape of the population is not assumed because it is unlikely that it would be known.

The elimination of unrealistic assumptions removes from the statistical model an element of unreality, thereby clarifying the relationship between the model and the empirical world. With a finite model, complete diagrammatic representation of every equally likely sample is possible. Even when such diagrams are not practical, they are still imaginable. Not only can the relationships between model and application be grasped more readily, but so also can the mathematical computations. Basing the computation of confidence intervals on a model for a finite,

49

discontinuous population makes it unnecessary to use calculus or other techniques for dealing with infinite, continuous distributions. A person with no advanced mathematics can have a complete understanding of the distribution-free procedures in this chapter.

Purposes of interval estimation

Assuming normality makes it easy for an investigator to decide which parameter to estimate. He estimates the mean or the standard deviation because these two parameters completely define the normal distribution. He does not need to decide whether to estimate the median or mode instead of the mean because an estimate of the mean of a normal distribution is necessarily also an estimate of the median and the mode. He does not have to worry about whether to estimate the range because the range is infinite. Nor does he need to choose between estimating the semi-interquartile range and the standard deviation because an estimate of the semi-interquartile range can be obtained from an estimate of the standard deviation. It is unnecessary to estimate the amount of skewness because a normal distribution, being symmetrical, has no skewness. In view of such facts it is no wonder that statisticians pay little attention to *reasons* for estimating population parameters; with the normality assumption the answer is simple: estimate the mean to get an estimate of the magnitude of the scores, and estimate the standard deviation to get an estimate of the variability.

When the investigator uses, instead of normal curve estimation procedures, estimation procedures that apply to a finite population of any shape, there are many different, independent parameters that he can estimate, and he must decide which ones are appropriate. For example, he can estimate either the mean or the median, or perhaps both, because these parameters do not necessarily have the same value. The freedom thus gained by the investigator through being released from his ties to the normal curve carries with it the responsibility for making decisions where previously there were no decisions to be made, such as deciding whether to estimate the mean or the median.

Discussions of interval estimation frequently are vague and sometimes are even misleading regarding the purpose of estimation of population parameters. It may be quite convenient to demonstrate a computational procedure by computing a confidence interval for the mean IQ in a given high school, but the reader has a right to know what conceivable practical importance such a confidence interval would have. Who would claim that the mean IQ in that particular high school is really of general interest? On the issue of the appropriate parameter to estimate, one might wonder if the mean is the parameter most likely to be of use in such a case. Perhaps some other parameter would be of more interest, such as the proportion of individuals with IQs less than 80. Now one might think that the confidence interval for the mean of a normally distributed population could provide such information, but that is not the case; the population standard deviation would also have to be estimated. Furthermore, if one were to try to reconstruct a normal population by estimating both the mean and the standard deviation simultaneously, he would need to use confidence intervals dealing with both parameters jointly in order to get an estimate of the proportion of individuals below some specified value. In other words, an interval estimate of a mean, even of a normal distribution, provides little of use unless one is interested in the mean itself, and it is hard to imagine why one would be interested in the total of the IQs in the high school divided by the number of students.

The preceding illustration is not intended as a blanket condemnation of the use of artificial examples in statistics. Indeed, hypothetical situations and unrealistically small samples may be quite useful in examples for providing an understanding of a statistical technique. But artificiality in an example should never be of a kind that promotes *misunderstanding* of a technique.

People have become so accustomed to seeing examples of useless confidence intervals that they fail to consider how a given confidence interval will be used. Perhaps this is partly the result of thinking in terms of infinite populations with parameters having more universal significance than those of the mundane populations actually sampled. Now it may be contended that in some instances a person simply wants a numerical value to look at, to give him ideas about the population. But if this is the desire, why compute a confidence interval? Why not simply use the sample mean alone?

Perhaps the most likely use for an estimate of a mean is in connection with totals. An estimate of a mean can be multiplied by the population size to get an estimate of a total. Confidence intervals for means can similarly be changed into confidence intervals for totals. For example, this would be done in estimating the total wheat production of a country or the total amount of tax that a city will collect. However, totals would not be of much interest in connection with IQs of high school students or with reading test scores for all adults in a city.

An estimate of the proportion of a population in a given category can be converted to an estimate of the total number within the population that fall within that category. This conversion permits such practical estimates of total frequencies as the number of potential purchasers of some marketable item or the number of mental defectives requiring institutional help.

Reasons for estimating the population median are the same as those for estimating any other population percentile, namely, to estimate the score below which there is some given percentage of the individuals in a population. If desired, this percentage can be converted to a frequency by multiplying by the population size. The application is similar to the application of the estimate of a proportion. The objective of an estimated proportion is the determination of the number of people in a given category, whereas the objective of estimating a median is to determine the cutting point which divides two quantitatively different categories, the lower half of the population and the upper half. It is not always the 50th percentile (the median) which is of major interest, of course. For example, consider the task of designing a typewriter for children. Some children's fingers are so weak that it would be virtually impossible to make a manual typewriter with keys they could depress. A realistic aim would be a typewriter requiring a hitting force that would be acceptable to 90 percent of 8- to 10-year-olds, in which case the 90th percentile and the confidence interval for the 90th percentile would be computed.

A common reason for estimating parameters is for drawing inferences about a more general population than the one actually sampled. For instance, a researcher may use a random sample of rats to find the median lethal dosage of a poison, although his interest is not in the median dosage for the population sampled, but for other rats as well. Similarly, other investigators may want to estimate the size of next year's high school class or next month's values on the stock market. In such cases the investigators may draw *statistical* inferences about the population actually sampled, but must generalize beyond this population on a *nonstatistical* basis. This, for example, would be necessary in the case of designing a typewriter for children because the children tested are not the ones to use the typewriters in the future ... and even if they were, the children would change in the meantime.

Besides being interested in the estimation of parameters of a single population, a person sometimes wants to estimate the difference between the parameters of a population under two conditions ("treatments"). For example, a random sample of cattle is taken to determine experimentally which of two types of feed provides the greatest weight gain. Since feeds are not equally expensive, accessible, easy to mix, etc., more than likely the investigator would want to find out not only which feed was best, but also how much more the entire herd would weigh if it were given that feed instead of the other. He wants an estimate of the mean difference in effectiveness of the two feeds because this estimate can be multiplied by the number of animals to provide an estimate of the total difference for the entire herd. Confidence intervals can be computed for this purpose.

Sometimes a parameter of a population is estimated in order to draw inferences about *individuals* within the population. For instance, a population regression line is estimated in order to draw inferences from the X measurement of an individual about his Y measurement or to make inferences in the opposite direction. The usefulness of a regression line resides in the ease in measuring one variable and the difficulty in getting measurements on the other variable because of time, money, etc. By using the regression line and a measure of X a better estimate of Y can be gained than from using only the mean Y value as the estimate.

Distribution-free confidence intervals

Confidence intervals are computed on the assumption of random sampling, and since only *finite discontinuous* populations can be randomly sampled, all valid confidence intervals can be presumed to be based on random samples from such populations. This is a more realistic starting point in a discussion of interval estimation than the assumption of an infinite, continuous distribution, but it complicates the discussion. In particular, it will be necessary to introduce some new terminology to be used in connection with confidence levels.

First, we will consider the conventional distinction made between *probability* and *confidence*. Any procedure for constructing confidence intervals must have a known probability, under the assumption of random sampling, of providing an interval that contains the parameter value being estimated. When such a procedure provides intervals that have a 95 percent *probability* of including the population parameter, there is said to be, for any interval obtained by the procedure, 95 percent *confidence* that it contains the population parameter, and similarly for other confidence levels. There are two changes in assigning levels of confidence for the distribution-free procedures to follow: (1) Conventional levels like .95 and .99 are not always possible, and so less conventional levels like .82 are sometimes necessary, and (2) even with the less conventional levels, one must, instead of assigning an exact level of confidence, indicate that the level of confidence is *at least* some specified value. We will now consider these two points.

The assumption of a normal distribution, which by definition must be infinite and continuous, allows confidence intervals to be established for any desired level of confidence, but confidence intervals for finite, discontinuous populations can be established for certain levels of confidence only. For example, any time a normal curve confidence interval can be computed, it is possible to compute a 99 percent confidence interval for the mean, whereas for samples of certain sizes from finite populations there may be no way to compute confidence intervals with

54 Statistical Inference: The Distribution-free Approach

exactly 99 percent probability of including the population mean. Only a finite number of confidence levels are possible for a given sample from a finite population, and the discrete distribution of possible confidence levels may not include 99 percent or any value that large.

Another consideration that makes normal curve confidence intervals more convenient to discuss is that it makes no difference whether the limits of the confidence interval are considered to be included within the interval or excluded from the interval. On the other hand, with confidence intervals for finite populations, only by introducing the somewhat unrealistic assumption of no ties in the population could one construct a confidence interval in the "exclusive" sense. For instance, given a 90 percent confidence interval, 34.2 to 38.9, for the mean of a finite population with no assumptions about ties in the population, there is 90 percent confidence that the population mean is *no less than* 34.2 and *no greater than* 38.9. A normal curve confidence interval with those values could be interpreted in the same way, or because the probability of a score of *exactly* 34.2 or 38.9 is infinitely small, it could be said that there is 90 percent confidence that the population mean is *greater than* 34.2 and *less than* 38.9.

The statement that a finite population with no assumptions about ties might permit a 90 percent confidence statement that the population mean is no less than 34.2 and no more than 38.9 is only approximately correct. More precisely, without making assumptions about ties in the population, it should be said that there is *at least* 90 percent confidence that the population mean is no less than 34.2 and no more than 38.9. Thus, there is an "*at least* 90 percent" confidence interval, not a 90 percent confidence interval. For instance, suppose there is a population of four scores and that the median of a population containing an even number of scores is defined as the value halfway between the two middle values. If all four scores are the same, the probability is 1 that the median is no greater than the larger of two randomly selected scores. If all four scores are different, the probability is 5/6 that the median is no greater than the larger of the two randomly drawn scores. This is shown in the following diagram, where the scores from small to large are represented by letters A, B, C, and D, in order of size.

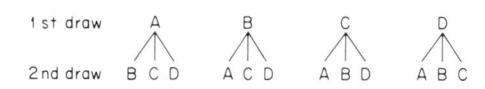

```
1st draw      A         B         C         D
             /|\       /|\       /|\       /|\
2nd draw    B C D     A C D     A B D     A B C
```

In ten of the twelve equally probable pairs a *C* or a *D* occurs and so the probability is 10/12, or 5/6, that the median is no greater than the larger of two randomly drawn scores. With two or three ties at the median value the probability is 1. Thus it can safely be said without making any assumptions about ties in the population that for any population of four scores the probability is *at least* 5/6 that the median is no greater than the larger of two randomly selected scores. One cannot, of course, say that the probability is *exactly* 5/6 because this requires the assumption of no ties in the population.

Distribution-free confidence interval for a median

The median is defined here in the conventional way: the value of the middle score when there is an odd number of scores or the value halfway between the values of the two middle scores in an even number of scores. Thus, in any population, no more than half of the scores are above the median value and no more than half below.

Example 4.1

Thompson wants to set up a one-sided confidence interval for the median of a population. He takes a random sample of five scores from a population of ten. Regardless of whether the population contains ties, the probability of the first drawn score being above the value of the median is no greater than 5/10. (It will be *less* than 5/10 in the case of two or more scores in the population having the same value as the value of the median and is *exactly* 5/10 without such ties.) Given that the first-drawn score is above the median, the probability that the second score is above the median is no greater than 4/9, because no more than four of the nine remaining equally probable scores could be above the median. If both the first- and second-drawn scores are above the median, the probability that the third is above the median is no more than 3/8, and the probability is no more than 2/7 and 1/6 for the fourth- and fifth-drawn scores. Thus, the probability that all five scores are above the population median is no greater than $5/10 \times 4/9 \times 3/8 \times 2/7 \times 1/6$ = .004 (approximately). There is then a probability of *at least* .996 that all five scores are not above the population median; that is, there is at least 99.6 percent confidence that the median is at least as large as the smallest score in the sample.

Example 4.2

Thompson could have computed a two-sided confidence interval in a similar fashion if he had wanted to. The probability of all five sample scores being *below* the population median is the same as that for all of them being *above* the median, namely, no greater than .004. The probability, then, that either all five sample scores are above the median or else that all five sample scores are below the median is no greater than .004 + .004 = .008. An "at least 99.2 percent" two-sided confidence interval for the population median is the inclusive range of the sample scores.

Published probability tables for the median (e.g., Meredith, 1967) are based on the assumption of an infinite continuous population. Therefore, the confidence intervals differ in several respects from those obtained by the exact procedure given here. In the first place, the "exclusive" confidence interval implied by the assumption of continuity of the population, of the form $X_a <$ mdn $< X_b$, should be changed to the form $X_a \leq$ mdn $\leq X_b$, in recognition of the discontinuity of any population of measurements. Second, also because of the actual discontinuity, the confidence level is not exact but is an "at least" level. Finally, the confidence levels associated with the intervals are different because of the assumption of an infinite population. This, however, simply makes the published confidence levels more conservative than necessary, because the procedure recommended here, wherein the actual size of the population is used, *increases* the confidence level for an interval. This is illustrated in the following example.

Example 4.3

In randomly sampling an infinite population, the probability is 1/2 that a score is not less than the median, regardless of the number of scores already drawn. In Example 4.1, if the population had been regarded as infinite, the probability of all five scores being equal to or greater than the median would be computed as $1/2 \times 1/2 \times 1/2 \times 1/2 \times 1/2 = 1/32$ or approximately .031, instead of .004. The one-sided confidence interval for the median being at least as large as the smallest score in the sample would then be regarded as an at least 96.9 percent confidence interval, instead of an at least 99.6 percent interval. The confidence level for any other one-sided or two-sided interval would be affected in the same way. The reason is easy to comprehend. For any sample, the fewer the unknown scores in the population, the more certain one can be regarding the location of the population median, the greatest certainty being when there are *no* unknown scores. Therefore, for any sample, the smaller the population from which it was drawn, the more certain one can be about the size of the population median. Consequently, probability tables for *infinite* populations are conservative, and the confidence levels could be raised by using the actual population size. Of course, when the population is large compared to the sample, there is little advantage in computing the confidence interval on the basis of the actual population size.

Distribution-free confidence interval for a proportion

Example 4.4

Suppose a random sample of two individuals is taken from a population of five individuals. The problem is to estimate the proportion of the population in category X. Designate the five individuals as A, B, C, D, and E, and represent the equally probable outcomes of taking two individuals, randomly without replacement, in this way:

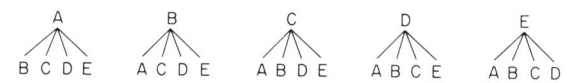

```
     A          B          C          D          E
    /|\        /|\        /|\        /|\        /|\
   B C D E    A C D E    A B D E    A B C E    A B C D
```

First assume that none of the individuals is classified as X. Then each of the 20 events shown in the diagram will give two individuals that are non-X. Thus, when the population proportion of X is 0, every sample of two individuals will have a proportion of 0.

Next assume that one of the individuals, say individual A, is X, and that the others are non-X; the population proportion of X is 1/5. Then every pair that contains A will have a sample proportion of X equal to 1/2, and every pair that does not contain A will have a sample proportion of 0. Eight of the twenty events contain A, and so the probability is 8/20 of getting a sample proportion of 1/2 and 12/20 of getting a sample proportion of 0, when the population proportion of X is 1/5.

Then assume that A and B are classified as X and the rest as non-X. The population proportion of X is 2/5. Then every pair that contains either A or B has a sample proportion of 1/2, any pair that contains both A and B has a proportion of 1, and those that have neither A nor B have sample proportions of 0. The probability of a sample proportion of 0 is 6/20, the probability of a sample proportion of 1/2 is 12/20, and the probability of a sample proportion of 1 is 2/20.

Now consider a population proportion of 3/5 in which A, B, and C are classified as X. There are only two of the twenty events that do not contain an A, a B, or a C, namely, D-E and E-D. Thus the probability of a sample proportion of 0 is 2/20. The probability of a sample proportion of 1/2 is 12/20, and the probability of a sample proportion of 1 is 6/20.

Suppose A, B, C, and D to be classified as X, the population proportion being 4/5. The probability of a sample proportion of 0 is 0; the probability of a sample proportion of 1/2 is 8/20; and the probability of a sample proportion of 1 is 12/20.

Finally, suppose that all five individuals are classified as X. Then the probability of a sample proportion of 0 is 0, the probability of a sample proportion of 1/2 is 0, and the probability of a sample proportion of 1 is 1.

58 Statistical Inference: The Distribution-free Approach

The following table summarizes the computed probabilities:

Sample proportion	1	0	0	2/20	6/20	12/20	1
	1/2	0	8/20	12/20	12/20	8/20	0
	0	1	12/20	6/20	2/20	0	0
		0	1/5	2/5	3/5	4/5	1
		Population proportion

From this table can be derived a second table in which the cell with the highest probability in each column is shaded:

Sample proportion	1	0	0	2/20	6/20	*12/20*	*1*
	1/2	0	8/20	*12/20*	*12/20*	8/20	0
	0	*1*	*12/20*	6/20	2/20	0	0
		0	1/5	2/5	3/5	4/5	1
		Population proportion

The sample proportion corresponding to the row in which a shaded cell falls is the most probable sample proportion for a sample taken from the population corresponding to the column in which the shaded cell falls. For instance, for a population proportion of 0 the most probable sample proportion is 0. For a population proportion of 3/5 the most probable sample proportion is 1/2. Since the smallest probability in the shaded cells is 12/20, or 60 percent, the probability is at least 60 percent that we will obtain a sample with a sample proportion whose row will intersect the column for the actual population proportion at a shaded cell. Thus there is at least 60 percent *confidence* that the actual population is a population whose proportion column has a shaded cell in the row of our obtained sample proportion. Therefore, if our obtained sample proportion is 0, we have 60 percent confidence that the population proportion is 0 or 1/5. If our sample proportion is 1/2, we have at least 60 percent confidence that the population proportion is 2/5 or 3/5. And if our sample proportion is 1, we have at least 60 percent confidence that the population proportion is 4/5 or 1.

Provided the procedure for constructing a confidence interval is determined independently of the sample results, preferably before the sample is taken, there are a number of alternative ways to construct confidence intervals for a proportion. We have shown a way to set up a 60 percent confidence interval for the population proportion based on a sample of two taken from a population of five individuals. We will now show how to construct a confidence interval with a higher level of confidence, for the same sampling situation. We shade the high probabilities in the table as follows:

	1	0	0	2/20	*6/20*	*12/20*	*1*
Sample proportion	1/2	0	*8/20*	*12/20*	*12/20*	*8/20*	0
	0	*1*	*12/20*	*6/20*	2/20	0	0
		0	1/5	2/5	3/5	4/5	1
		Population proportion					

Now for each of the populations the probability is at least 18/20, or 90 percent, that a sample proportion will be in one of the rows containing the shaded cells for that population. If we obtain a sample proportion of 0, we have at least 90 percent confidence that the population proportion is 0, 1/5, or 2/5. If we obtain a sample proportion of 1/2, we have at least 90 percent confidence that the population proportion is 1/5, 2/5, 3/5, or 4/5, which is trivial because the population proportion is certain to be one of those proportions. If we obtain a sample proportion of 1, we can make an at least 90 percent confidence statement that the population proportion is 3/5, 4/5, or 1.

Distribution-free confidence interval for a mean

Example 4.5

Brown wants to construct a one-sided confidence interval for the mean of a population of 22 scores, from which he has randomly selected 19 scores. The 19 scores in the sample are arranged below, in order of size:

11 11 13 14 15 15 16 16 16 17 18 19 20 21 23 25 27 27 29

What degree of confidence can Brown have that the three remaining scores do not contain a score larger than 29? First, consider a population of 22 scores with a single largest score (not several scores having the largest value). The odds are 19:3 that the largest score will be in a random sample of 19 scores rather than among the three undrawn scores, and so the probability is 19/22 that the largest score will be in the sample. Although more complicated, a determination of the same probability value could be arrived at by means of the application of the probability tree described in Example 2.7, in which the undrawn scores are regarded as being randomly selected not to be drawn. Then we consider the probability of three undrawn scores *not* containing the largest score in a population of 22 to be the same as the probability of a randomly selected group of three scores not containing the largest score. In a probability tree each of the 22 scores is represented in the first row and each of these is connected to the other 21 scores in the second row, which in turn are each connected to the other 20 scores in the third row. The proportion of the 9,240 equally probable branches that do not contain the largest score is the probability of not getting the largest score in the first row, not getting it again in the second row, and not getting it in the third row, which is 21/22 × 20/21 × 19/20, which is 19/22. Where there are two or more scores in the population tied for the largest value, the probability is *greater* than 19/22 that *at least one* of these values is in the first 19 randomly drawn scores. Therefore, without making any assumptions about the presence or absence of ties for the largest score (or any other score, for that matter), the probability is *at least* 19/22, or approximately .864, that there are no larger scores among the three undrawn scores than are contained in the sample of 19 scores. Brown has, then, at least 86.4 percent confidence that none of the scores in the population is greater than 29, which is the largest score in the sample. If none of the scores in the population is greater than 29, then the population *total* is not greater than the sum of the 19 scores plus three 29s, which is 440. Brown has 86.4 percent confidence therefore that the *total* of the population is not greater than 440, and he has 86.4 percent confidence that the *mean* is not greater than 440 divided by 22, which is 20.

Example 4.6

Jones has been assigned the task of constructing a confidence interval for the mean score of 10 items. Since underestimation and overestimation are both costly, Jones must construct a two-sided confidence interval. He has a random sample of 8 of the 10 scores to work with:

19 22 25 26 27 28 31 34

What degree of confidence can Jones have that neither of the two remaining scores is smaller than 19 or larger than 34? For populations of 10 scores where there is only one largest score and only one smallest score, the probability is exactly 8/10 that the ninth score to be drawn is neither the largest nor the smallest score in the population, since it is equally probable that the ninth score is any one of the 10 scores and since eight of the 10 scores are neither the largest nor the smallest score. Given that the ninth score is neither the largest nor the smallest score, the probability is 7/9 that the tenth score also is neither the largest nor the smallest score. The probability therefore is 8/10 $\times$ 7/9 $=$ 56/90, or approximately 62.2 percent, that neither of the two undrawn scores is either the largest or the smallest score in the population; in other words, the probability is 62.2 percent that a sample will contain both the largest and the smallest scores. For populations where there is more than one largest or smallest score, the probability is greater than 62.2 percent that a sample contains at least one of the largest scores and at least one of the smallest scores. Therefore, with no assumptions regarding ties, the probability is *at least* 62.2 percent that there are no larger and no smaller scores in the population than are contained in the sample. The eight scores in Jones's sample add up to 212, and when Jones adds two 19s to this total he gets 250 as the lower limit of his confidence interval for the total of the 10 scores. The upper limit for the total is 212 plus two 34s, which is 280. The 62.2 percent confidence interval for the *total* is 250 to 280, and so the 62.2 percent confidence interval for the *mean* is 25 to 28.

Distribution-free confidence interval for a difference between means

There are occasions when it would be useful to have a confidence interval for a difference between population *totals*. It is traditional, however, to compute confidence intervals for a difference between *means*. The following discussion will consider confidence intervals for both parameters since the confidence intervals are closely related.

Example 4.7

Harris wants to compute a one-sided confidence interval for the difference $\mu_B - \mu_A$. More specifically, he is interested in the maximum amount by which the mean of *B* can be expected to exceed the mean of *A*. Harris draws, at random, the following scores from populations *A* and *B*:

A: 4 9 12 12 18 18 19 20

B: 5 7 13 17 19 24 25 25 27 28 31

62 Statistical Inference: The Distribution-free Approach

The eight scores in sample A were taken from a population of 10 scores and the 11 scores in sample B were taken from a population of 12 scores. What sort of confidence can be assigned to the statement that the two undrawn scores in population A are at least as large as 4 and also that the undrawn score in population B is no larger than 31? Using the procedure of Example 4.5, Harris finds that the probability is at least 8/10 that there are no smaller scores in the population than the smallest in a random sample of eight scores taken from a population of 10 scores, and that the probability is at least 11/12 that there are no larger scores in the population than the largest in a random sample of 11 scores taken from a population of 12 scores. The joint probability, then, that a random sample of eight scores from population A contains as *small* a score as there is in the population and that a random sample of 11 scores from population B contains as *large* a score as there is in population B is at least 8/10 × 11/12, which is approximately 73.3 percent. Harris can, therefore, state with at least 73.3 percent confidence that *neither* of the two undrawn scores in population A is *less* than 4 and that the undrawn score in population B is *no larger* than 31. The 11 scores in sample B add up to 221 and when Harris adds a 31 to this he gets 252. The eight scores in sample A add up to 112 and when Harris adds two 4s he gets 120. There is 73.3 percent confidence that the population B *total* minus the population A *total* does not exceed 252 − 120 = 132. There is 73.3 percent confidence that the population B *mean* minus the population A *mean* does not exceed 252/12 − 120/10 = 9.

Example 4.8

Suppose Harris wanted to compute a *two-sided* confidence interval for $\mu_B - \mu_A$ based on the data given in Example 4.7. What confidence value could be assigned to the statement that the two undrawn scores in population A fall within the sample A limits and also that the undrawn score in population B falls within the sample B limits? In Example 4.6 it was shown that the probability is at least 56/90 that the last two scores in a population of 10 scores from which eight have been randomly drawn will fall within the sample range. What is the probability that the last score in a population of 12 scores falls within the range of the first 11 scores? If there were one smallest and one largest score in the population, the probability would be 10/12 that the last score is neither the largest nor the smallest score and therefore falls within the sample range. If there were more than one smallest or more than one largest score, the probability would be greater than 10/12. Therefore, the probability is at least 10/12 that after 11 scores have been randomly taken from a population of 12 scores, the remaining score falls within the range of the 11 sample values. The joint probability, then, that a random sample of eight scores from population A contains scores that are as large as any in the population and scores that are as small as any in the population, and that the same is true with regard to population B for a random sample of 11 scores from population B, is at least 56/90 × 10/12 = approximately 51.9 percent. Harris can state with 51.9 percent confidence that the two undrawn scores in population A fall within the sample A limits, 4 to 20, and that the undrawn score in population B falls within the sample B limits, 5 to 31. The next step is to see what this confidence statement implies about differences between population totals. It implies that the population B total minus the population A total is *no less* than 226 (the population B total if the undrawn score were 5) minus 152 (the population A total if the two undrawn scores were 20s), which is 74, and that the difference is *no greater* than 252 (the population B total if the remaining score were 31) minus 120 (the population A total if the remaining scores were 4s), which is 132. There is 51.9 percent confidence that the population B *total* minus the population A *total* is between 74 and 132. There is 51.9 percent confidence that the population B *mean* minus the population A *mean* is between (226/12 — 152/10) and (252/12 — 120/10); that is, between 3.6 and 9.

Distribution-free confidence interval for an individual

The general impression given in discussions of statistics is that interval estimation concerns estimation of population parameters only. In the previous discussion of distribution-free confidence intervals for a mean, however, it was necessary to determine the probability that an individual or several individuals outside the sample have measurements within the range of the sample measurements. This indicates that a confidence interval could have been computed for the undrawn individual or the mean of the undrawn individuals. In fact, normal curve techniques also permit making confidence statements about individuals. Regression procedures are sometimes used for this purpose. There are times, however, when one would like to make inferences about individuals, and regression procedures cannot be used.

Sometimes it is desirable to make an inference from a sample to an individual or individuals in the same population when the statistical assumptions of regression are not satisfied or when regression cannot be used because there are no correlated variables on which to base the estimate. The possibility of such inferences will be shown here by the explicit presentation of a distribution-free procedure already implied in the discussion of distribution-free confidence intervals for the mean.

Before considering the distribution-free procedure let us see what purpose is served by a confidence interval for an individual. Why compute a confidence interval for an individual rather than simply measure the individual? The answer is that it is sometimes not practical to measure the individual. A situation of this kind is that in which the testing of items weakens or destroys them. For example, a severe test of an electronic component of a guided missile may render the tested item unusable, yet it may well be worthwhile to test a number of such components in order to accurately estimate the quality of the one that *will* be used.

A distribution-free method can be used to establish confidence intervals for the next individual to be drawn from a population. Suppose we take a random sample of 19 items from a population containing 20 or more items in order to obtain a 95 percent one-sided confidence interval for the 20th item. First we arrange the 19 measurements from smallest to largest, for example, 34.3, 36.8, ..., 45.7, where 34.3 is the smallest measurement and 45.7 is the largest. It can be asserted with 95 percent confidence that the 20th item will have a measurement equal to or larger than 34.3, because when 20 measurements are randomly drawn from a population, any one of the measurements is as probable as any other to be drawn last. Thus, if all 20 measurements are different in size, the probability that the last-drawn measurement (the measurement not in the sample of 19) will be the smallest is 1/20, or .05; therefore the probability that the 20th of a sample of 20 different measurements will *not* be the smallest is .95. When there are two or more measurements among the 20 measurements that are tied for the smallest value, the probability that the undrawn measurement is not smaller than the smallest in a sample of 19, is, of course, 1. Without assumptions about the presence or absence of ties, therefore, the probability is *at least* .95 that the undrawn measurement is no smaller than the smallest score in the sample. Since it is realistic to allow for the possibility of ties, we should in this example say that we have *at least* 95 percent confidence that the 20th measurement will be equal to or greater than 34.3.

For a 99 percent confidence statement it is necessary to have a sample of at least 99 measurements. For a sample of 99 measurements we can assert with at least 99 percent confidence that the next (the 100th) measurement will be no smaller than the smallest measurement in the sample of 99. Now consider a 95 percent confidence interval for the 100th measurement. There is at least 95 percent confidence that the

100th measurement will be no smaller than the fifth smallest measurement in the sample of 99 because there is a probability no greater than 5/100 that the 100th measurement will be one of the lowest five measurements.

Two-sided distribution-free confidence intervals for individuals can be constructed if desired. For instance, for a sample of 24 measurements there is at least 92 percent confidence that the 25th measurement will *not* fall outside the range of the first 24 measurements because the probability that the 25th measurement *will* fall outside the range of the first 24 measurements is no greater than 2/25, or 8 percent. Shorter confidence intervals can be determined in a similar fashion. There is, for instance, at least 84 percent confidence that the 25th measurement will not fall outside the range of the *next-to-smallest* and the *next-to-largest* measurements (that is, the range of the middle 22 measurements) in the sample of 24 because the probability that the 25th measurement will be one of the two smallest or two largest measurements is no greater than 4/25 or 16 percent.

The relative efficiency of distribution-free estimators for the mean and the median

In normal curve estimation the standard error of the mean is less than the standard error of the median, and so it is more efficient to estimate the population median (which in a normal distribution would also be the mean) by using a confidence interval for the mean than by using a confidence interval for the median. Also, in every particular instance, a computed confidence interval for any specified level will be shorter (and therefore more precise) for the mean than for the median. Now in distribution-free estimation, as was pointed out earlier, the mean and the median of the population are not assumed to be the same value, and so it is not desirable to use the sample mean as an estimate of the population median or the sample median as an estimate of the population mean. Nonetheless, the following example involving a comparison of distribution-free confidence intervals for the mean and the median is instructive inasmuch as it shows a relationship between confidence intervals that could not exist in normal curve estimation.

Example 4.9

In Example 4.6, we computed an at least 62.2 percent confidence interval for the mean to be 25 to 28. If, in that instance, we computed a confidence interval for the *median,* we could specify a *100* percent confidence interval for the median as 25 to 28. The median could not be less than 25 even if both remaining scores were less than the smallest of the sample scores, and the median could not be greater than 28 even if both remaining scores were larger than all of the sample scores. This example is not intended to suggest that in every instance the significance level for an interval for the median of the same length as the interval for the mean will be higher than that for the mean. It simply shows that it is possible in some cases.

Normal curve confidence interval for a proportion

Normal curve procedures for constructing a confidence interval for a proportion are sometimes given. The distribution of scores in a population makes no difference when the only relevant information about the distribution is the proportion of individuals within a category; thus, no assumptions regarding the shape of the population can allow more precise estimates of a population proportion than the distribution-free procedures that were given. The shape of the theoretical sampling distribution of sample proportions is dependent on the population proportion, being symmetrical about the population proportion when it is .5, positively skewed for population proportions below .5, and negatively skewed for population proportions above .5. Recognition of this changing shape of the theoretical sampling distribution is customarily displayed by reference to the inappropriateness of normal curve confidence intervals when the *sample* proportions are extreme, because the confidence interval for the population proportion is centered on the obtained sample proportion.

Normal curve confidence interval for a difference between totals

The distribution-free procedure for computing a confidence interval for a difference between means required, as a preliminary step, constructing an interval for a difference between totals. This, of course, is not required for normal curve confidence intervals for a mean, but if differences between totals have the practical value they appear to have, it would be useful to be able to compute such intervals with the normal curve approach, when the normality assumption is appropriate.

Example 4.10

Suppose we have computed a 95 percent confidence interval for a difference between means by the use of normal curve techniques.

Can we directly convert this interval, for example, 14.2 to 19.3, into an interval for a difference between totals? The answer is that we can certainly do this if the two finite populations are the same size. For instance, if both populations consisted of 100 scores, we could convert our 95 percent confidence interval of 14.2 to 19.3 into a 95 percent confidence interval for a difference between totals of 1,420 to 1,930. This conversion can be based on the *differences* between means without consideration of the individual population means. The inference that two population means differ by 14.2 points when both populations contain 100 scores is logically equivalent to the inference that the population totals differ by 1,420 points, because when the larger population mean is multiplied by 100 to give the population total, it will give a total that is 14.2 × 100 points larger than multiplying the smaller population mean by 100. Similarly for the inference that two population means differ by 19.3 points. A 95 percent confidence interval for a difference between means of 14.2 to 19.3 therefore logically implies a 95 percent confidence interval for a difference between totals of 1,420 to 1,930.

Example 4.11

When the population sizes differ, the confidence interval for a difference between means *cannot* be converted into a confidence interval for a difference between totals by the procedure used in Example 4.10. Suppose that population A contains 100 scores and that population B contains 200 scores. The 95 percent confidence interval for a difference between means $\mu_B - \mu_A$ is 14.2 to 19.3. This interval cannot be converted to a confidence interval for the difference between totals because of the difference in the size of the populations. That population B has a mean that is 14.2 points greater than the population A mean does not imply the amount by which the totals differ. If the population B mean were 20.2 and the population A mean were 6, the difference between totals would be 4,040 − 600 = 3,440, since population B contains 200 scores and population A 100 scores. But for the same amount of difference between means (14.2), if the population B mean were 40.2 and the population A mean were 26, the difference between the totals would not be 3,440, but would be 8,040 − 2,600 = 5,440. Thus, when the population sizes differ, an estimate of a difference between means cannot be converted to an estimate of a difference between totals because not only the size of the difference between means but also the size of the individual means affects the difference between totals.

Example 4.12

Although one cannot obtain a normal curve confidence interval for a difference between totals for populations of different sizes directly from a normal curve confidence interval for a difference between means, it can be obtained in a manner similar to that for the distribution-free situation. Suppose that for population A the 95 percent confidence interval for the mean is 46.5 to 50.4 and for population B the 95 percent confidence interval for the mean is 48.8 to 51.6. There is 90.25 percent (that is, 95 percent $\times$ 95 percent) confidence that both population means fall within their 95 percent confidence intervals. Now suppose population A has 100 scores and population B has 200. Then the 95 percent confidence interval for the total for population A is 4,650 to 5,040, and the 95 percent confidence interval for the total for population B is 9,760 to 10,320. The two-sided 90.25 percent confidence interval for population B total minus population A total is 4,720 to 5,670, the 4,720 being obtained by subtracting 5,040 from 9,760 and the 5,670 being obtained by subtracting 4,650 from 10,320.

It is likely that Example 4.10, in which the sample sizes were the same, is more typical of practical situations in which a confidence interval for a difference between totals is desired. It is doubtful that there are very many occasions in which one would be interested in a difference between totals of populations of different sizes. On the other hand, suppose a person wanted to estimate the difference in the total weight of a herd of cattle for two alternative feeds. A random sample from the herd would be given feed A and another random sample feed B. The confidence interval for a difference between totals would refer to the difference between the total weight of the herd if they all were given feed A and the total weight if they all were given feed B. The "population" sizes would be the same because they would be "populations" of measurements of the same group of animals under two different treatments.

Normal curve confidence interval for an individual

Distribution-free confidence intervals for measurements of individuals have been considered earlier. In the present discussion the concern is with *normal curve* confidence intervals for the next individual to be randomly drawn from the same population. It is convenient to consider the procedure of establishing a confidence interval for the next individual as a special case of establishing a confidence interval for the *mean* of the next sample, when the next sample consists of only one measurement.

The discussion of estimation of the measurement of an individual could begin with a sample taken from a normal population with unknown mean but known variance, but it seems unrealistic to suppose that the variance would be known if the mean were unknown, and so it will be assumed that both the mean and variance are unknown; therefore, the *t* distribution will be used.

Very little modification of the standard procedure for establishing a confidence interval for a difference between population means is needed. Where $\hat{\sigma}$ is the square root of an unbiased estimate of the population variance based entirely on the variability within the samples,

$$\frac{(\overline{X}_1 - \overline{X}_2) - (\mu_1 - \mu_2)}{\hat{\sigma}\sqrt{1/n_1 + 1/n_2}}$$

is distributed as *t* with the number of degrees of freedom associated with $\hat{\sigma}$. When both samples are taken from the same population, $\mu_1 = \mu_2$, and so

$$\frac{\overline{X}_1 - \overline{X}_2}{\hat{\sigma}\sqrt{1/n_1 + 1/n_2}}$$

is distributed as *t*. A two-sided confidence interval for $\overline{X}_2$ can then be computed. For example, the probability is .95 that

$$\overline{X}_2 = \overline{X}_1 \pm (t_{.05})(\hat{\sigma})\sqrt{\frac{1}{n_1} + \frac{1}{n_2}}$$

where $t_{.05}$ is the value of *t*, with the degrees of freedom possessed by $\hat{\sigma}$, that is required for two-tailed significance at the .05 level. For a confidence interval for a difference between population means, one uses

$$\hat{\sigma} = \sqrt{\frac{\Sigma x_1^2 + \Sigma x_2^2}{n_1 + n_2 - 2}}$$

with both $\hat{\sigma}$ and *t* having $(n_1 + n_2 - 2)$ degrees of freedom. When we want to estimate the second sample mean before we draw the second

70 Statistical Inference: The Distribution-free Approach

sample, however, the estimate of the population standard deviation must be based entirely on the variability within the first sample, and so we use the estimate,

$$\hat{\sigma} = \sqrt{\frac{\Sigma x_1^2}{n_1 - 1}}$$

which has $n_1 - 1$ degrees of freedom.

Example 4.13

Consider a numerical example of the use of data from a single sample to establish a confidence interval for the next individual to be drawn from the same population. This is a special case of sample-mean-to-sample-mean estimation, the case when the second sample consists of only one measurement.

The Navy wants to estimate the strength of a rubber life raft in terms of how much air pressure it can withstand. The test of a life raft weakens it so much that the tested raft is not usable. The potential cost in human lives of a defective life raft is great enough to make it worthwhile to test several of them in order to precisely estimate the strength of one to be issued. Navy officials take a random sample of six rafts and test each. The mean pressure the six rafts can stand is 82.6 units of pressure and the estimate of the population standard deviation,

$$\sqrt{\frac{\Sigma x_1^2}{n_1 - 1}}$$

is 5.3. Assuming the population to be normally distributed, what is the 99.9 percent confidence interval for the strength of the next raft taken at random from the stockpile? The two-sided 99.9 percent confidence interval for the next (the seventh) raft is

$$X_7 = 82.6 \pm (6.859)(5.3)\sqrt{\frac{1}{6} + \frac{1}{1}}$$

where 6.859 is the value of t with five degrees of freedom required for two-tailed significance at the .001 level.

Example 4.14

The previous example showed how to construct a two-sided confidence interval when one is desired, but in the hypothesized situation a one-sided confidence interval would be more appropriate because the objective is to determine whether the next life raft will be strong enough; it could be inadequate by being too weak but not by being too strong. Few tables of t show the required value for one-tailed significance at the .001 level; in this example, therefore, a 99.5 percent one-sided confidence interval will be constructed instead of a 99.9 percent one-sided confidence interval. There is 99.5 percent confidence that the next raft can withstand a pressure of at least

$$82.6 - (4.032)(5.3)\sqrt{\frac{1}{6} + \frac{1}{1}}$$

or at least 59.5. 4.032 is the value of t with five degrees of freedom required for one-tailed significance at the .005 level.

Confidence interval for an individual based on linear-regression procedures

Linear-regression procedures can improve the estimate of the measurement of an individual by using a measurement of a different variable which is correlated with the one being predicted. Thus, instead of simply estimating the measurement before the next individual is selected, the individual is selected and a measurement on a related variable is made to predict the measurement on the variable of interest. But if we have already selected the individual, why not directly measure the characteristic we are interested in instead of a related characteristic? At first glance it would appear to be a game of guessing one measurement from knowledge of another, but there are practical uses for the technique. For example, a person might be able to make some measurement of a life raft that would permit accurate estimation of its breaking strength, without weakening the raft. Whenever one can obtain measurements of a related characteristic that permit him to estimate another characteristic which it is impractical to directly measure, this technique may be useful.

Linear-regression techniques assume more than simply the normality of variables X and Y. They assume a random sampling of a *bivariate normal distribution* of paired measurements. In a bivariate normal distribution, X is normally distributed, Y is normally distributed, the distribution of Y for every X value is normal, and the distribution of X for every Y value is normal. Also, all X distributions have the same standard deviation and all Y distributions have the same standard deviation.

Only two parameters are required to completely describe a normal distribution: μ and σ. For a bivariate normal distribution, five parameters are required to completely describe the distribution: μ_X, μ_Y, σ_X, σ_Y, and r_{XY}. Typically all of the parameters of the bivariate population are unknown. Estimation of population regression lines requires working with sample estimates of each of these parameters.

There is inconsistency among reference books with regard to the *standard error of estimate*. There is disagreement about both the definition of the standard error of estimate and the proper formula for a particular definition. Within a given sample, the regression line for predicting Y from X minimizes the sum of the squared deviations of Y values about the regression line, and thereby minimizes the standard deviation of the Y values about the regression line. In using the regression line instead of the actual plotted points in estimating Y from a value of X, the error in estimating is the difference between the regression line (estimated value) and the actual value of Y. Some statisticians regard the standard error of estimate as simply the standard deviation of the sample measurements about the regression line. The formula for this standard error of estimate would be

$$\sqrt{\frac{\Sigma(y - y')^2}{n}}$$

where y is the actual value, y' is the estimated value, and n is the number of pairs of measurements. This formula is appropriate as a starting point in setting up a confidence interval for a measurement estimated from a value of X, and if the measurements in the sample are almost normally distributed about the regression line, one can use normal curve tables to set up a confidence interval. From the practical standpoint, however, one might wonder why anyone would be interested in such estimation. After all, why take the X value of an individual in the *sample* and estimate his Y value by means of the regression line? All individuals in the sample would already have been measured on variable Y as well as X since that was necessary in order to compute the correlation coefficient and construct the regression line. Why estimate a value that is already known and recorded? Yet regression is frequently discussed in this way. This is unfortunate because such discussions obscure the practical utility of linear-regression procedures. Realistically, one would expect to use a regression line to estimate the value of Y for another individual *not* in the sample upon which the regression line is based but taken at random from the same population. For this type of estimation, there are two additional sources of error that are not involved in "estimating" measurements within the sample: the population standard deviation about the population regression line may be different from the sample standard deviation about the sample regression line, and the population regression line may not coincide with the sample regression line. The proper formula for the standard error of estimate of an individual not in the sample but in the same population, using the sample regression line, is (Walker and Lev, 1953):

$$\sqrt{\frac{\Sigma(y-y')^2}{n-2}}$$

The division by $n-2$ instead of n in the expression increases the size of the standard error of estimate, which is appropriate because of the additional sources of error in estimating beyond the sample individuals.

The normal distribution assumption

There can be no doubt of the great value of normal curve estimation procedures when the normality assumption is appropriate. For any level of confidence, a much more precise confidence interval can be computed by normal curve procedures than by distribution-free procedures. Thus, when normal curve estimation procedures are appropriate, they should be used because of the increased precision of estimation. By the same token, when normal curve procedures are *inappropriate*, they should not be used because the *spuriously precise* confidence intervals will frequently be in error.

It would be foolish to suggest that the normality assumption is *never* appropriate, just as it would be to suggest that it is *always* appropriate. Each population must be judged individually. Yet there are some general arguments for and against the assumption of normality that are likely to be helpful in arriving at a decision in a particular instance. Some of these arguments will be presented here.

The central limit theorem argument for normality

The most common theoretical reason, independent of consideration of the variable being measured, for the normality assumption is the central limit theorem. The central limit theorem has been mathematically derived and proved. It states that any population with a finite variance (and every population that exists, and that can therefore be randomly sampled, does have a finite variance) will, when randomly sampled, provide sample means that approach normality. The larger the

74 Statistical Inference: The Distribution-free Approach

sample size, the more nearly normal the distribution of sample means. There is more to the theorem, and the measure of closeness to normality requires a mathematical definition, but this part of the theorem is adequate for our purposes.

The central limit theorem was developed on the assumption of an infinite population, but the implication that sample means tend to be more nearly normally distributed than the original scores in the parent population also holds true for the type of sampling we have considered to be typical in actual practice: random sampling of a finite population, without replacement.

Example 4.15

Consider the following frequency distribution on the left. This is a distribution representing a population of ten scores. Suppose we take a sample of two scores, without replacement, from this population and compute their mean. The distribution of equally probable means is shown on the right.

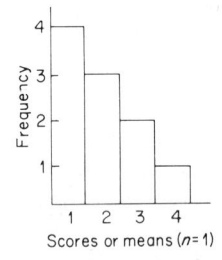

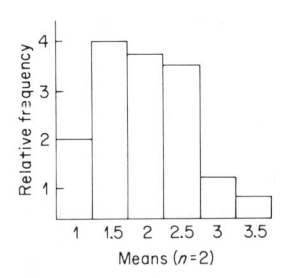

It can be seen that even though we started off with an extremely skewed distribution, and took only small samples (two scores each), the distribution of means is almost symmetrical. It may be wondered why we are interested in the normality of the sampling distribution of means instead of the normality of the distribution of individual scores in the population. The reason is that normal curve confidence intervals are based on the normality of the population only to the extent that this affects the normality of the sampling distribution of means.

According to the central limit theorem, as we increase the sample size, the distribution of sample means becomes more nearly normal. For any sample size, the more nearly normal the population, the more nearly normal the sampling distribution of means. When the sample size is small and the population is far from normal, the distribution of sample means will be less normal than when larger samples are taken from distributions closer to the normal distribution. Thus, we see that the central limit theorem is most useful when it is used in conjunction with assumptions about the shape of the population.

The statement that the distribution of sample means becomes more normal as the sample size increases must be qualified when referring to finite populations that are sampled without replacement. There is certainly a tendency for the sample means to cluster more closely about the population mean as the sample size increases, but in some cases the *shape* of the sampling distribution of means is closest to normal when the sample size is half of the population size; as the sample size deviates from this middle size in either direction, the sampling distribution of means becomes less symmetrical.

For instance, in the preceding example, the frequency distribution of the population is the same as the theoretical sampling distribution of means of samples of one score each. Because the population contains 10 scores, for every sample of one score there is an *undrawn sample* of nine scores. Since the sum of the population scores is 20, when the drawn score is 1, the sum of the nine undrawn scores is 19 and the mean is 2 1/9. When the drawn score is 2, the mean of the undrawn scores is 2. When the drawn score is 3, the mean of the undrawn scores is 1 8/9. When the drawn score is 4, the mean of the undrawn scores is 1 7/9. Therefore, a histogram of the sampling distribution of means of the undrawn nine scores, with the values 1 7/9, 1 8/9, 2, and 2 1/9, along the base, would show *frequencies* of 1, 2, 3, and 4, respectively. (Do not confuse the *scores* and the *frequencies* in the original population frequency distribution.) That is, the frequency distribution would be the mirror image of the distribution for means of samples of one score each. Since the sampling distribution of means of nine *undrawn* scores is the same as the sampling distribution of means of nine *drawn* scores, we see that, so far as shape of the distribution is concerned, the distribution of means of samples of nine scores deviates from normality as much as the distribution of means of samples of one score, because the distributions are mirror images. Obviously, the same mirroring of distributions exists for the sampling distributions of means for samples of two scores and for samples of eight scores, for three-score samples and seven-score samples, and so on, for sample sizes equally distant from a sample size of 5.

Since normal curve estimation procedures would rarely be used with samples that are as large as half of the population size, the preceding discussion is of theoretical importance only.

The binomial distribution argument for normality

The *symmetrical binomial distribution*, in which $p = q = 1/2$, approaches normality rapidly as the sample size increases. In fact, the normal distribution has been shown mathematically to be the limiting form of the symmetrical binomial distribution as the number of trials increases. If a characteristic within a group of individuals can be considered to be genetically determined by a number of independent elements, each of which is as likely to be possessed as not, there is implied a symmetrical binomial distribution of the number of elements possessed by the individuals resulting from the genetic process. And, provided each element additively contributes the same amount as any other element to the magnitude of the measured characteristic, we have a symmetrical binomial distribution of the measurements.

The argument based on the binomial distribution refers to a *process* which generates a binomial distribution, not to a binomial distribution of characteristics for a randomly sampled population. A random sample of individuals from an existing population is *regarded* as a random sample from an infinite population of individuals produced by the binomial process. If the population actually randomly sampled cannot be regarded as typical of the individuals produced by the postulated process, then we cannot regard a random sample from that population as a "random sample" of the infinite population of individuals produced by the process. For example, if we considered the heights of humans in general to be produced by such a process, we would not be justified in regarding the height of a random sample of basketball players as a random sample of the infinite population produced by the binomial process and would therefore require some other argument if we were to assume normality of the population sampled. Since nobody considers a randomly sampled population to possess a perfect binomial distribution of measurements of height, weight, or any other characteristic, we will in the following example treat the argument as it is usually presented, in terms of sampling an infinite binomial population generated by a binomial process. For the computation of probability values this is equivalent to sampling a finite binomial population *with* replacement.

Example 4.16

With large samples, even when the probability of occurrence and nonoccurrence of an event are unequal, the distribution of the number of events tends to symmetry. Consider, for instance, a population in which 40 percent of the elements are non-*A* and 60 percent are *A*.

The asymmetry of the population is obvious in the following distribution on the left. Notice, however, the almost symmetrical distribution on the right, which is the theoretical sampling distribution of the number of A individuals in samples of five individuals each, taken with replacement from the population.

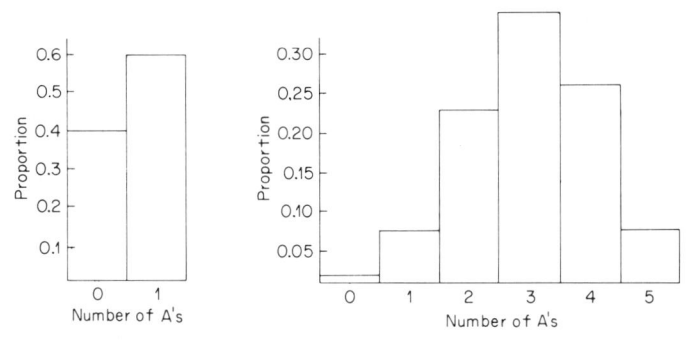

Assume that every A element contributes one unit of characteristic X and that every non-A element makes no contribution. In that case the distribution on the right would be the distribution of the amount of characteristic X.

Nonmathematical arguments for and against normality

It can be assumed that for certain characteristics there is an optimum amount of that characteristic. Animals possessing considerably less or considerably more than the optimum amount have a small chance of survival. The chances of survival diminish as one goes further and further in either direction from the optimum amount of the characteristic, and so one can expect a concentration of measurements somewhere around the middle of the distribution with a gradual tapering off toward the ends of the distribution. This argument is stronger when the characteristic is genetically determined because natural selection can, over a number of generations, accumulate the slight effects within the individual generations.

The shape of the sample distribution is sometimes examined to decide whether the assumption of population normality is tenable. However, a positively skewed sample could have come from a normal population and a normally distributed sample could have come from a positively skewed population. Since sample distributions seldom are of precisely the same shape as the population distribution, it is necessary to judge whether the shape of a sample distribution is so nonnormal that one should conclude that the sample did not come from a normally distributed population. Presumably, if one tests his sample for normality and rejects the hypothesis of a normal population, he will not use normal curve statistical procedures. But what happens if the hypothesis of normality is not rejected? A person would presume that failure to reject the null hypothesis would not permit him to conclude that the population is normal because that would be "accepting the null hypothesis"; nonetheless, many statisticians either imply or explicitly state that it would then be appropriate to assume normality. Researchers who employ large samples may feel disciminated against, because rejection of the hypothesis of normality is more likely to occur with large samples than with small, because increasing the sample size makes the test more sensitive.

Decidedly nonnormal distributions are sometimes to be expected. For instance, U-shaped distributions, where the measurements cluster at the ends rather than in the middle of the distribution, have been observed. This kind of distribution might result when there is a very high intercorrelation among the items on a test. A very homogeneous test could yield scores that tended to be either near zero or else near a perfect score simply because persons who knew the answers to a few key questions could answer almost all the rest of the questions, while those who did not know the answer to those questions could get almost none of the other questions. A U-shaped distribution also could reflect the existence of forces that prevented persons from taking a moderate stand on some issue, as indicated, for example, by responses to an attitude questionnaire. In an attitude questionnaire persons who answered one question in a certain way might tend to answer a cluster of other questions in a similar way, while persons who answered that question differently might be inclined to answer the others differently, too. Distributions of cloud coverage over England are U-shaped, showing that almost complete cloud coverage and almost completely clear skies are much more frequent than intermediate cloud coverage.

Positively skewed distributions can be expected when a task is hard for almost all of the persons performing it and there is a lower limit to the performance scores. The scores pile up near the lower limit, such as a score of 0 on a test, although a few people may get considerably higher scores. When many scores pile up at a point near 0, there cannot be symmetry if some persons get scores considerably higher than 0.

Negatively skewed distributions result from similar tendencies associated with an *upper* limit. A positively skewed distribution for one measure frequently implies a negatively skewed distribution for an inversely related measure. For example, if we were to measure the time to run a mile, the median time of a group of persons might be 5 or 6 minutes, but a few persons would take 10 minutes. On the other hand, probably no person in the group would run the mile in much less than 5 minutes, and so we might have a somewhat positively skewed distribution of times. On the other hand, if we were to have the persons run for 4 minutes, several would run almost a mile, but a few would run less than half a mile, and so there would probably be a negatively skewed distribution of *distances.* Not only do physiological limits act as upper limits against which scores pile up, but maximum scores on relatively easy tests do also.

Transformations and normality

Mathematical transformations are sometimes carried out when the assumption of normality is more plausible for the population of transformed measurements than for the population of original measurements. Any transformation that makes a distribution more normal must be a nonlinear transformation because a linear transformation has no effect on the shape of a distribution. This is an important consideration because the mean is invariant only under linear transformations (Stevens, 1951). In other words, if and only if a transformation is linear can one determine from the mean alone of a set of measurements the value of the mean of the set of transformed measurements or determine from the mean alone of the set of transformed measurements the mean of the original measurements. For instance, if the mean of two quadrupled numbers is 24, we know that the mean of the numbers is 6, but if we were told that the mean of the squares of two numbers is 25, we would not be able to determine the mean of the original numbers from that information alone. To consider only positive integral values, the original numbers could be 5 and 5 or they could be 1 and 7. Because of this lack of invariance of the mean under nonlinear transformations, a confidence interval for the mean of nonlinearly transformed measurements does not provide us with the necessary information to construct a confidence interval for the mean of the original measurements.

Thus, only nonlinear transformations will serve the function of changing the shape of the distribution and in doing so they make it impossible to construct confidence intervals on the mean of the original measurements. Consequently, if we are interested in the mean of the original measurements, transformations to normalize the distribution should not be made because doing so would make it impossible to construct a confidence interval for the mean of the population of original measurements.

In practical applications where confidence intervals for means are of interest primarily because of their implications for totals, the inappropriateness of nonlinear transformations is obvious. For example, one can transform the weights of a sample of animals by taking their logarithms and use these to construct a confidence interval for the mean logarithm of the weights of the population of animals. But we cannot multiply an estimated mean logarithm by the number of individuals in the population to get an estimate of the logarithm of the population total. Multiplying the estimated mean logarithm by the population size gives an estimated logarithm of the *product* of the weights of the population of animals.

Example 4.17

A numerical example showing the relationship between a set of numbers and their logarithms will help clarify the preceding discussion. Suppose, for instance, we have three numbers: 2, 4, and 8. The corresponding common logarithms of the numbers are 0.3010, 0.6021, and 0.9031. The mean of these logarithms is 0.6021. The antilogarithm of this value is 4, which is the *geometric mean* of the original numbers. If we multiply 0.6021 by 3 to get the total of the logarithms, we obtain 1.8063. This value is the logarithm of 64, which is the *product* of our three original numbers. Thus, the total obtained by multiplying the mean of the logarithms by the number of numbers is the logarithm of the *product* of the numbers, not the logarithm of the *sum* of the numbers.

Although a confidence interval for the product of the weights of animals in a population might be useful, it is hard to imagine practical situations where this is so. Generally, we want to estimate the *sum* of the weights of all of the animals if we are going to ship them, sell them by the pound, etc.

For constructing a normal curve confidence interval for the next *individual* from the same population, a nonlinear transformation to normalize the distribution of measurements has no adverse effect at all on the interpretation. For example, the 95 percent confidence interval for the *logarithm of the weight* of the next individual can be directly converted into a 95 percent confidence interval for the *weight* of the next individual by taking the antilogarithms of the limits of the confidence interval. For example, if the 95 percent confidence interval for the logarithm of the weight in pounds of the next individual is 0.3010 to 0.6021, the 95 percent confidence interval for the weight in pounds of the next individual is 2 to 4. Although we said earlier that the mean is invariant under a linear transformation only, we must qualify that statement to read "The mean of *more than one number* is invariant under a linear transformation only." In the special case of a single measurement, the mean can be invariant under nonlinear transformations. The logarithmic transformation is one such nonlinear transformation. A cubic transformation is another. For instance, if we know that 27 is the mean of a single cubed number (the mean, of course, being simply the cubed number itself), we know that the mean of the original number (the mean being the original number itself) is 3, even though the transformation is a nonlinear transformation. This invariance does not hold for the mean of more than one number; we cannot determine from the mean alone of two or more cubed numbers the value of the mean of the numbers that were cubed.

The bivariate normal distribution assumption

There are indications that the assumption of a bivariate normal distribution is not taken seriously. One indication is the repeated reference to correlations within a restricted range of one or both of the correlated variables to show how a restriction of the range of scores reduces the correlation and thereby makes prediction from a regression line less precise. Such demonstrations are invalid because linear regression cannot be appropriate for both the entire population and the restricted population. If the entire population has a bivariate normal distribution, the restricted population cannot have. Any mathematical argument to show the extent to which restriction of the range reduces the accuracy of estimation should take this fact into consideration.

The arguments for a normal distribution are certainly relevant here because a bivariate normal distribution requires that the X and Y variables both be normally distributed. However, as was pointed out earlier, a bivariate normal distribution is more than simply a joint distribution of two normally distributed variables, and so some additional arguments are required. For example, an argument for linearity is desirable, because a bivariate normal distribution must show a linear relationship between variables X and Y inasmuch as the column means fall in a straight line and the row means fall in a straight line. In the following discussion, we will briefly consider the possibility of developing arguments for a bivariate normal distribution analogous to those for a normal distribution.

First, consider the central limit theorem. The counterpart of this theorem for bivariate distributions would appear to be a theorem which stated that any bivariate distribution, when randomly sampled, will provide *sample centroids* (centers of gravity) that approach a bivariate normal distribution. For example, for samples of 2, for every pair of points on a scattergram, we find a point midway between them. The theorem would imply that these midpoints would have a distribution closer to a bivariate normal distribution than the original points on the scattergram. The larger the sample size, the closer the distribution of sample centroids will be to a bivariate normal distribution.

Perhaps modern genetic theory could provide arguments for a bivariate normal distribution based on pairs of binomial distributions. Suppose characteristic A to be distributed by a binomial genetic process, and characteristic B also to be distributed by a binomial genetic process. Then if one assumed independence of the two processes, a large number of A elements, and a large number of B elements, a theoretical joint distribution of the number of A elements and the number of B elements possessed by individuals should be very close to a bivariate normal distribution, possessing zero correlation between variables A and B.

An earlier argument for measurements piling up around the middle of a distribution was that animals possessing considerably more or considerably less than the optimum amount of some characteristic have a small chance of survival. Now a joint distribution of two such characteristics should show a clustering of measurements around the center of the joint distribution if the two characteristics had independent effects on survival.

The nature of the sample bivariate distribution is of course one factor to consider in deciding whether the assumption of a bivariate normal distribution is reasonable. Presumably, goodness-of-fit tests could be performed to see whether to reject the hypothesis of a bivariate normal population.

The minimal requirement for a bivariate normal distribution is that each of the variables be normally distributed, and so skewness or other lack of normality in either variable makes the bivariate normal assumption untenable.

Even when one is satisfied that the relationship between two variables is sufficiently linear to justify linear-regression procedures, he may be able to improve his accuracy of estimation by transforming the measurements to provide a more linear relationship between them. For example, correlations between height and weight are commonly used to illustrate linear correlation. The sample distributions certainly do not suggest much nonlinearity of relationship between height and weight in adults. There is a temptation to regard this correlation as readily explicable, but it is more complex than it first appears. The complication is that there is a high positive correlation between height and weight over a group of adults, but virtually no correlation between height and weight within individuals during adult life. During adulthood a person's height is virtually constant, whereas his weight may fluctuate considerably. Since weight varies but height does not, there is no correlation between height and weight within individual adults during adulthood.

In children, on the other hand, one would expect increases in height within an individual to accompany increases in weight. And we might be justified in expecting a relationship between height and weight over children of various ages to exhibit the type of relationship that would be found in a longitudinal study of an individual. Consequently, we would *not* expect the height and weight of children of various ages to be *linearly* related, because equal changes in a child's height do *not* go with equal changes in weight, which is what a linear relationship implies. Since weight is a function of the volume, one might expect that (height)^3 would be more nearly linearly related to weight. Actually, tables of norms for children (Almy, 1955) suggest that weight changes directly with the *square* of the height, not the *cube* of the height. (Does this mean that older children are slimmer, leggier, or what?)

If one correlated (height)^2 with weight, he could use linear regression in setting up confidence intervals for the next individual to be selected from the population. For instance, a confidence interval for height expressed in (inches)^2 could be converted to a confidence interval for height in inches by simply taking the square root of the confidence-interval limits.

Validity of measurements

The arguments given in the chapter on sampling to justify regarding the lottery as the ideal physical random sampling procedure concerned pellets drawn from a container. The pellets represented individuals who, after being selected, had to be measured. No reference was made to the assumption of *validity* of the measurements and, in fact, none is necessary if the inferences are restricted to measurements obtained by the same procedure and measuring devices. For instance, a meaningful confidence interval for the mean weight of a population can be based on distorted scales that systematically give readings that are 2 percent too high. The estimation would be meaningful but would have to be restricted to measurements made by those particular scales. Biased measurement (not biased sampling) can provide the basis for statistical inference, then, with regard to measurements obtained by the same procedure and the same measuring devices without knowledge of the type or amount of bias. Needless to say, in most practical situations one would like to draw inferences about measurements by other measuring devices and therefore would like to know the degree of bias.

Reliability of measurements

We will now consider two assumptions that are essential for both normal curve and distribution-free estimation procedures, essential in fact for any statistical estimation procedure. These assumptions concern the stability or reliability of the measurements.

First, consider how instability of measurements over time affects the interpretation of confidence intervals. Suppose we wanted to construct a confidence interval for the mean weight of a group of individuals and that the span of time over which the measurements were made was long enough for the individuals to increase substantially in weight. Some persons would say that a perfectly valid confidence interval could be computed in such a case, provided the population had been randomly sampled. But what would such a confidence interval refer to: the population mean weight at what time? After all, the population mean weight changes, and so there is no single mean weight to be estimated, but any one of a number of mean weights for various times. If it were possible for us to simultaneously measure all of the sample individuals, we would say that we have a confidence interval for the mean weight at the time of measurement. When, however, we have various times of measurement and when the population mean is different for the various times, the picture of the population mean we are trying to estimate is blurred. We must not have blind faith in the ability of random sampling to pull us through such difficulties; instead, we must require that confidence intervals based on random sampling provide unambiguous numerical values and this would seem to imply that confidence intervals must refer to population parameters at some particular time.

If we are estimating a characteristic that is stable over the times of measurement, we can construct a meaningful confidence interval for the population mean even though our measurements are spread over time. The confidence interval would refer to the population mean at the time of measurement, or, more precisely, to the population mean at any time during the time span of the measurements.

Besides stability of measurements over the measurement period, we also need stability of measurements of individuals over all possible samples in which they could have appeared. To illustrate this point, consider a situation in which we want to estimate the mean performance of a population. Suppose the performance of each individual was stable over the time of the sample measurement period, but varied according to the composition of the sample; when certain individuals were in the sample to which he belonged he would perform better than otherwise. If we constructed a confidence interval for the population mean performance, to what would it refer? Certainly not to the mean performance at the time of sampling, because there is no single mean performance at any particular time. At any given time, there is the mean performance of the population when the individuals are split in a particular way into groups for measurement, and other values of mean performance at that time for other ways of splitting the population into groups for measurement. Nor would there be any basis for estimating the mean performance during the time of measurement for the entire population measured one at a time, independently of other individuals, because the sample used for estimating does not provide such isolated measurements for the sample individuals. In order to have a confidence interval for a mean refer to something that is empirically meaningful, it appears necessary for the population mean under estimate to be the one that would be obtained if each individual could be measured independently of every other.

If our sample measurements of individuals are the same as would have been obtained if each sample individual were the only member of the sample, we can then construct meaningful and useful confidence intervals. The intervals would refer to the population mean at the time of measurement, the population mean that has the same value no matter how the population is split into groups for measuring.

We thus have two ways in which measurements must be reliable in order to permit empirically meaningful confidence intervals: (1) Each individual provides the same measurement as if he had been measured at any other time during the sample measurement time period, and (2) each individual provides the same measurement as if he were the only individual in the sample.

If both of these assumptions had to be met completely, there would be few opportunities for using confidence intervals, but they do not have to be met completely. Like the assumption of normality, these assumptions present conditions that it is hoped will be closely approximated. If the conditions are closely approximated, there is little error in making inferences as though they were met exactly. We will now consider factors that affect the closeness with which these assumptions are met: sources of unreliability. We will classify sources of unreliability into three categories: the measuring instrument, the person doing the measuring, and the individual being measured.

As we stated earlier, a biased measuring device may, by reason of the error being constant, permit valid statistical inferences about the measurements obtained by that device. When the errors are not constant, however, the unreliability makes the validity of statistical inferences questionable. The measurement properties of an instrument may change as a function of the magnitude of the measurements from the previously measured individuals. For a durable, simple measuring instrument like a ruler the magnitude of the preceding measurements should have little effect. But if the instrument is complex or has been misused, as in overloading an ammeter or voltmeter, a distortion may result. What about the effect of the number of previous measurements on the measuring instrument? In complex electronic equipment continued use may result in an imbalance and consequent misreading unless a periodic compensating adjustment is made. This is, to some extent, the effect of change in the properties of the circuit components as they warm up and can be reduced somewhat by making sure the equipment has had ample time to warm up. There are, however, some components which are involved only during the measuring process, and a series of measurements may affect these. What about fluctuations over time that are not a function of previous measurements? Long-term changes can be ruled out since our concern is with the short period of time over which our sample is measured. Provided we have isolated our measuring devices from known sources of distortion such as electrical interference from passing cars or temperature changes resulting from people opening and closing the door, there should be little trouble of this sort.

The person doing the measuring may also contribute to measurement fluctuation. In the operation of some pieces of measuring equipment, the operator must make an adjustment based on sensory perception such as an adjustment to equalize the brightness of two lights or the loudness of two sounds. In such cases, there is the possibility of perception being distorted by the previous measurement; for example, a moderate light following an intense one appears dimmer than otherwise. Also, as a result of the number of measurements the person may become fatigued or less attentive, and this can contribute to variation in the measurements. In addition, as a function of physiological changes independent of previous measurements, there are changes in a person's perceptual acuity from moment to moment and changes in his alertness.

Finally, we must not overlook the fact that the characteristic being measured may be quite variable during the time over which the measurements are made. In measuring the size of protozoa with short life spans, the measuring time may be long enough to allow the later measured protozoa to grow considerably. Usually, of course, we are not interested in characteristics that change so rapidly over a short time span, but it is easy to overlook the variation that may occur in attitudes and other mental characteristics during the measuring time, easy to overlook the fact that the characteristics being measured do fluctuate. This may be in part due to the tendency of researchers to regard intelligence, personality, and many other characteristics as stable and to refer to fluctuations in the measurements as the result of "error." Such a view is useful for some purposes, but some phenomena are better understood if we concentrate on the changeableness of psychological characteristics over short periods of time.

Inferences about measurable individuals

In the previous section it was shown that the distribution-free procedures discussed in this chapter, like all other interval estimation procedures, are based on the assumption of reliability of measurements over the measurement period and are not appropriate unless this is a reasonable assumption.

Examination of the distribution-free procedures will also show that they deal only with measurements that are used in estimation and so do not take account of the possibility that some of the sample individuals may not provide any measurements at all, as for example persons who refuse to be interviewed. This is also true of normal curve estimation procedures, of course. Does this fact indicate that interval estimation procedures are appropriate only when it is reasonable to assume that all individuals in the population would provide usable measurements if they were selected in a sample? If the answer were "yes," there would be very few instances where interval estimation would be appropriate, because it is reasonable to assume that generally a fair number and sometimes even a large number of individuals in the population would *not* provide usable measurements. The answer to the question is "no," however; the procedures of interval estimation are still valid even when sizable proportions of the population would be unmeasurable, provided the confidence statements are restricted to the population of measurable individuals within the sampled population.

In the chapter on sampling, mention was made of questionnaire studies in which a number of questionnaires are never returned. Since persons who return questionnaires may differ systematically in relevant ways from those who do not return questionnaires, it was indicated that one should in such cases restrict his inferences to the population of individuals within the sampled population who would return the questionnaires. Actually, the existence of unmeasurable individuals in the sample just calls attention to something likely to be overlooked when all of the sample individuals *are* measurable: that *all* inferences must be restricted to measurable individuals in the population, whether there are unmeasurable individuals in the *sample* or not, because there may be unmeasurable individuals in the *population*.

It seems odd that a person cannot make inferences about the randomly sampled population as a whole when his sample contains only measurable individuals. To see why statistical inferences cannot be made about the entire population even when all the individuals in a sample are measurable, let us consider an example. Suppose there are quite a few unmeasurable individuals in a population and that if we had been able to measure them, all of these individuals would have provided lower measurements than the measurable individuals. What happens if we make a 95 percent confidence statement about the mean of the entire population every time our sample contains only measurable individuals? Our estimation procedure is clearly biased toward overestimation. We would *not* have a probability of 95 percent of being correct in our statements about the population mean.

Also it is obvious that we must restrict our inferences in a questionnaire study not just to those individuals who would have returned the questionnaire, but to those who would have returned *usable* questionnaires. In other words, our confidence statements refer to the population of individuals within our sampled population who would have provided measurements that would have been used in making an interval estimation.

Example 4.18

Wilson randomly sampled the population of adults in his state, taking a sample of 100 names. He sent questionnaires to the 100 persons in his sample and got 80 usable returns. Of the 80 usable returns, 56 (that is, 70 percent) were in favor of Taxation Bill No. 45. He used a normal curve technique to compute a 95 percent confidence interval for the percentage in favor of the bill: 60 to 80 percent. Wilson has 95 percent confidence that between 60 and 80 percent of the people in the population who would provide usable responses are in favor of Taxation Bill No. 45. Furthermore, Wilson believes that between 60 and 80 percent of the entire sampled population are in favor of the bill because he sees no reason to believe that the unmeasurable individuals would have responded differently from the measurable ones. A friend, Black, believes that those persons who do not return acceptable questionnaires are more likely to be against the bill than those who do, and so he considers that one can be confident that the percentage of people in the entire population in favor of Bill No. 45 is no greater than 80 percent, but may well be somewhat less than 60 percent. Another friend, Taylor, believes, on the other hand, that persons who would not have returned questionnaires are more likely to be *for* the bill, and so he thinks one can be confident that at least 60 percent and perhaps considerably more than 80 percent of the people in the entire population are in favor of Taxation Bill No. 45. It can be seen that if the confidence interval for the population of individuals who would provide usable returns is published, readers can generalize to the entire sampled population on the basis of their judgment of differences between the measurable individuals and the unmeasurable individuals. Inferences about the measurable individuals are statistical inferences. Inferences about the entire sampled population are nonstatistical inferences, and must be supported by empirical argument, since there is no probabilistic basis for generalizing beyond the population of measurable individuals.

The population of individuals within the sampled population who would have supplied measurements to be used in making an interval estimation may be smaller than the population of individuals who could be available for measurement. For example, there may be some limitations of the measuring instrument itself to consider. If a gauge can provide no measurements greater than 50, all statistical inferences must be restricted to those individuals that could be measured by the gauge. The person who is convinced that there are few if any individuals in the population who would not be measurable with the gauge can, on the basis of his reasoning, generalize to the entire population, while other persons may prefer to consider the obtained estimates as possible underestimations.

Statistical and nonstatistical inferences

The terms *statistical inference* and *nonstatistical inference* have not been defined yet because each context in which they were used made the distinction between the two types of inference clear enough for the purpose.

Statistical inferences concern only the measurable individuals in a randomly sampled population. Measurable individuals are those individuals who would, if they were contained in the sample, provide measurements that would be usable in computing a confidence interval.

Statistical inferences concern measurable individuals for the time the sample measurements are made and the conditions under which they are made.

Nonstatistical inferences concern individuals for whom statistical inferences cannot be made or they concern the measurable individuals at other times or for other conditions than those under which the sample measurements are made.

Earlier we considered the type of nonstatistical inference in which inferences are made about the entire randomly sampled population. This type of inference is usually considered statistical, but we have seen that there is no probabilistic basis for it, and so it is nonstatistical.

Statistical and nonstatistical inferences have quite different logical foundations. Statistical inference is rigorous and independent of the investigator, whereas nonstatistical inference is based on qualitative arguments which, because they concern empirical events rather than mathematical logic, may convince one person but not another. The justification for a statistical inference resides in the computational procedure and the employment of a random sampling procedure, whereas the justification for a nonstatistical inference resides in empirical arguments based on the knowledge and beliefs of the investigator. Statistical inference is not necessarily *better* than nonstatistical inference but it differs from nonstatistical inference and its validity is judged in a different way, which is a more precise, standard way. Although nonstatistical inferences are necessary in order to make practical use of statistical inferences, there is value in making a distinction between the two types of inferences. A clear concept of statistical inference cannot be gained without distinguishing between generalizations which the statistical model permits and those which cannot be based on the model.

Estimation in psychological testing

Random sampling is required for confidence-interval estimation or in fact even to justify the use of a sample mean as an unbiased estimate of a population mean. Since in practice random sampling is rarely performed in the field of psychological testing, statistical estimation procedures have little relevance to *applied psychological testing,* although they apparently play a prominent role in *psychological testing theory.*

Both in the theory and the practice of psychological testing there are procedures which are called random sampling procedures, but these procedures are not lottery procedures or anything similar. In applied psychological testing, "random sampling" frequently means the systematic selection of individuals who are regarded as representative of some population. In psychological testing theory the notion of random sampling has even less similarity to the lottery model. For example, test items are sometimes regarded as random samples taken from an infinite domain of test items as if they were randomly taken from the brain of the person who constructed them. Also, the test score of an individual is (explicitly or implicitly) regarded as randomly selected from an infinite population of scores, the mean of which is called the "true score" of the individual.

Test norms are rarely based on random samples. They are usually derived from groups of readily available individuals who are regarded as representative of some larger group. Expert judgment, however, is no substitute for random sampling in ensuring "representativeness" and no statistical inferences can be made about individuals not in the norm group.

Even if test norms were based on random samples, statistical inferences could only be made about other individuals randomly selected from the same population, and the desired inferences do not usually concern randomly selected individuals. Instead, investigators want to make inferences about individuals who may be expected to differ systematically from other individuals in the population, such as inferences about individuals who go to counselors or other advisers for help.

To suggest that the statistics in applied psychological testing are descriptive statistics is not to imply that they are not complex. Quite complex techniques are required to obtain factor loadings, for example, but in the absence of random sampling they are simply complex descriptive measures of intercorrelation. As such, factor loadings are in the descriptive-statistics category along with reliability and validity coefficients.

The wide employment of complex statistical methodology in *psychological testing theory* could easily mislead the unwary into thinking that in *applied psychological testing,* inferences about other tests, other individuals, or the same individuals in the future have the same sort of statistical or probabilistic basis as inferences about a randomly sampled population. A good rule to follow in interpreting any so-called statistical inference is to see what population, if any, was actually randomly sampled. In applied psychological testing there is seldom a randomly sampled population, and so statistical inferences are seldom justified.

Chapter Five

Randomization Tests for Experiments

The previous chapter presented some distribution-free confidence-interval estimation procedures based on the lottery model of random sampling. The populations were finite and no assumptions were made about their shape or about the existence of tied measurements. These simplified procedures were found useful in providing insight into a number of aspects of both distribution-free and normal curve confidence-interval estimation. The inefficiency of these procedures, however, militates against their use as substitutes for normal curve procedures when the assumption of normality is justified.

This chapter will continue examining the application of the lottery model of random sampling in statistical inference. The inferences will concern hypotheses about experimental treatment effects. Once again the populations to be considered are finite and no assumptions are made about their shape or the existence of tied measurements. These procedures, because of their simplicity, reveal features of hypothesis testing that have generally gone unnoticed. Also, unlike the distribution-free interval estimation procedures, these distribution-free statistical tests are likely to be of practical value in a number of situations, because these tests are just as powerful (that is, sensitive) as their normal curve counterparts. Thus, they deserve consideration as potential alternatives to normal curve tests even when the assumption of normality is justified.

Random assignment and randomization tests

The distribution-free tests to be discussed rely on the random assignment of individuals to treatments. Random assignment is assignment based on random sampling of the group of individuals to be assigned. For example, the first ten subjects whose names are drawn are assigned to treatment A and the rest to treatment B. Random assignment therefore is based on lottery sampling of a "population" consisting of the individuals that will serve in the experiment. Random assignment cannot be performed in nonexperimental studies, and so the discussion of the tests in this chapter is restricted to experimental situations.

In the physical sciences, experimentation meant, and to a large extent still means, careful direct control over extraneous variables, precise manipulation of variables, and precise measurement of effects. The notion of random assignment to treatments was introduced into experimentation because in the biological and social sciences there is considerable variation from individual to individual. At first, random assignment was used in these fields to "equate" the individuals for the treatments, to ensure the absence of a systematic difference between subjects assigned to different treatments. In other words, the original objective of random assignment was to counteract the possibility of biased assignment of individuals to treatments. Sir Ronald Fisher advocated this use of random assignment a number of years ago. After using it for this purpose for some time, he realized that in addition to serving as a means of eliminating bias, random assignment also could provide the basis for a new class of statistical tests. These tests, called *randomization tests* or *permutation tests,* can be carried out without assumptions about the population shape whenever there has been random assignment of individuals to experimental treatments. Kempthorne (1952, 1955) has used the rationale underlying randomization tests as the basis for the development of experimental designs.

In the previous chapter it was necessary to develop new procedures of distribution-free estimation in order to get rid of the unrealistic assumption of an infinite, continuous population. It is unnecessary in this chapter to develop new procedures for testing experimental hypotheses because randomization tests have the desired properties. Randomization tests are distribution-free because they make no assumptions about population shapes; they concern finite populations with no assumptions about tied measurements; and the probability values are computed from random sampling models. In short, randomization tests are the hypothesis-testing counterparts of the distribution-free estimation procedures given in the previous chapter. Many properties of randomization tests correspond to those of distribution-free estimation procedures.

Levels of significance and probability

We earlier discussed the relationship between levels of confidence and probability to show the probabilistic basis of levels of confidence. There is a similar relationship between levels of significance and probability. If one uses a procedure that has a probability of .05 of rejecting the null hypothesis when the null hypothesis is actually true, a rejection of the null hypothesis obtained by such a procedure is said to be *significant* at the .05 level, and similarly for other levels of significance.

In allowing for the possibility of ties, we cannot specify a procedure that has a probability of .05 or .01 of rejecting the null hypothesis when it is true. We can, however, establish procedures that have probabilities *no greater than* .05 or *no greater than* .01 of rejecting the null hypothesis when it is true.

A randomization test derives a sampling distribution of a statistic (e.g., a correlation coefficient or a difference between means) from computations of the statistic for various divisions of the measurements. The type of random assignment determines the appropriate divisions. The results are significant at a given level if the proportion of the statistics in the sampling distribution that either equal the obtained value or exceed it is no larger than the significance level.

The null hypothesis for randomization tests

We want a null hypothesis for which a procedure can give a specified probability of rejecting the null hypothesis when the null hypothesis is actually true. The particular null hypothesis that has this property varies from one type of randomization test to another, but in general the null hypothesis for any application of a randomization test in an experiment can be stated in this form: "The effects of the treatments being compared are the same." This means that if any of the other treatments had been given instead of the one actually given, at the time it was given, the obtained measurement would have been the same. Now let us be more specific. The null hypothesis is always in terms of individuals and has this general form: "Every individual would have responded the same to any of the other treatments if it had been administered at his treatment time." The response referred to here is the one under investigation; we *do not* imply that *all* responses of the individual would be unaffected by treatment differences. In other words, the null hypothesis refers to identity of *measurement values* under the various treatments.

When we reject a null hypothesis it will be in favor of its negation. Instead of using the same null hypothesis for both one-tailed and two-tailed tests we will therefore require two different null hypotheses. The null hypothesis discussed in the preceding paragraph is the two-tailed null hypothesis and its rejection implies this two-tailed alternative hypothesis: "Some (one or more) of the individuals would have responded differently if one of the other treatments had been given at his treatment time." The one-tailed null hypothesis used when treatment *A* is expected to be more effective than treatment *B* is this: "No individual would have responded more strongly to the administration of *A* at his treatment time than to *B*." This null hypothesis would be rejected in favor of the following one-tailed alternative hypothesis: "Some (one or more) of the individuals would have responded more strongly to the administration of *A* at his treatment time than to *B*."

The widespread use of nonrandom samples in psychological experiments

A person conducting a poll may be able to enumerate the population to be sampled and select a random sample by a lottery procedure, but an experimental psychologist has not enough time, money, or information to take a random sample of the population of the world in order to make statistical inferences about people in general. Few psychological experiments use randomly selected subjects, and those that do usually concern populations so specific as to be of little interest. Whenever human subjects for psychological experiments are randomly selected, they are often drawn from a population of students attending a certain college and enrolled in a particular class who are willing to serve as subjects. Animal psychologists do not even pretend to take random samples, although they also use standard hypothesis testing procedures and reject null hypotheses about populations. These well-known facts are mentioned here as a reminder of the rareness of random samples in psychological experiments and of the specificity of the populations on those occasions when random samples are taken.

Statistical inferences cannot be made concerning populations that have not been randomly sampled; therefore, few experiments would be published if it were necessary to show that the experiment permitted a statistical inference concerning an important population, a population of general interest to the readers. We will now argue that random sampling of a population is not relevant to most psychological experimentation and that the lack of a random sample does not prevent drawing useful *statistical inferences*—about the experimental subjects actually used.

The irrelevance of random sampling for psychological experiments

In basic psychological experimentation, the principal interest in statistical hypothesis testing is to distinguish between treatment effects and effects arising from individual differences. The notion of population comes into the statistical analysis not because the experimenter has randomly sampled some population to which he wishes to generalize but because the only way he has been taught to interpret the results of a statistical test is in terms of inferences about populations.

There is, however, an alternative interpretation that can be given statistical results which does not involve the concept of a population. This interpretation is applicable to nonrandom samples or samples that have been systematically rather than randomly selected. The alternative approach will not permit statistical inferences about individuals not in the experiment but will permit statistical inferences about the experimental subjects. Statistical inferences about experimental subjects are justified by the random assignment of subjects to treatments, which is done by randomly sampling the "population" of subjects actually used in the experiment (Kempthorne, 1952 and 1955; Campbell, 1957; Campbell and Stanley, 1966; and Edgington, 1966).

The fact that statistical inferences can be drawn about the experimental subjects only is not as restrictive as it first seems. After all, to extend *statistical* inferences beyond the experimental subjects requires a random sampling of the population to which you wish to generalize, and this in turn requires a *finite population*. But experiments of a basic nature are not designed to find out something about a particular finite, existing population, but something about individuals already dead and individuals not yet born as well as those who are alive at the present time. If we were concerned only with an existing population we would have extremely transitory scientific laws because every minute individuals are born and individuals die, producing a continual change in the existing population. This is true not only of man and other animals but also of populations of plants. Since we cannot *statistically* generalize to a nonexisting population, we use a procedure of generalization that is fundamental to scientific investigation: nonstatistical generalization. We generalize from our experimental subjects to individuals who are quite similar with regard to those characteristics that we consider relevant.

For example, if the effects of a particular experimental treatment depend mainly on physiological functions that are almost unaffected by the social or physical environment, we might draw inferences about persons from cultures other than the culture of our subjects. On the other hand, if the experimental effects are easily modified by social conditions, we would be more cautious in generalizing to other cultures. In any event, the main burden of generalizing from experiments has always been carried by nonstatistical generalization, and this fact should be brought into the open.

This chapter will consider procedures of experimental design and data analysis from the standpoint of their usefulness for making statistical inferences about experimental subjects and nonstatistical inferences about other individuals.

Randomization test for a difference between independent samples

Example 5.1

First, let us consider an experiment in which we want to compare the effectiveness of two treatments, and can take only one measurement from each subject. Our object is to find out whether, with regard to the dependent variable measured, our treatments have the same effect. Obviously, we would not need a test to see whether the *obtained* measurements for both treatments were the same on the average, but this is not what we are interested in. We want to find out whether the measurements for one of the treatments are the same as they would have been if the other treatment had been given to those individuals. Take as an example an experiment with two treatments A and B, where we have predicted that subjects will perform better under A than under B. We use the first 10 available subjects or 10 specially selected subjects, making no attempt at random sampling. We have 10 treatment times and we assign a time to each subject, for example, by finding out when each subject is available. Then names are drawn from a hat to assign five subjects (with their prearranged treatment times) to treatment A and five to treatment B. The null hypothesis is that the performances are independent of the treatments; that is, that the performance of each subject is the same under one treatment as it would have been under the other treatment at that time. Under the null hypothesis, any difference between the performance under the two treatments is due solely to subject differences. Since the subjects were randomly assigned to the two treatments, if the null hypothesis is true, large differences between treatments can be attributed to "randomization error," discrepancies resulting from the random assignment of subjects. To test the null hypothesis, then, we consider the distribution of differences between the treatments under every equally probable division of the 10 performance measurements into two sets of five measurements. The number of ways 10 things can be divided into two groups of five things each is

$$\frac{10!}{(5!)(5!)} = 252$$

Suppose all five *A* subjects performed better than any of the *B* subjects. Only one of the 252 divisions assigns the five high-performing subjects to treatment *A* and the five lowest-performing subjects to treatment *B*. With a one-tailed (directional) test, we would have results significant at the .01 level, for example, in favor of the hypothesis that for the experimental subjects, performance was affected by the treatment. The probability under the null hypothesis of getting such a large difference in the predicted direction by random assignment alone is only 1/252. The conclusion that the treatments have different effects on our experimental subjects is justified by the random assignment of subjects to treatments, no matter how the subjects are selected.

At this point, we should be more specific regarding the hypothesis to be accepted in the preceding experiment. In saying that we conclude that for the experimental subjects performance was affected by treatment, we mean simply that some (one or more) of the subjects, at the time they were tested, would have responded differently if the other treatment had been administered. The reason that we must not draw conclusions about the treatment effect on *all* of the subjects is that rejection of the null hypothesis that every subject responded the same as he would have under the other treatment simply permits accepting the negation of this hypothesis, the negation being that at least one subject would have responded differently.

Now in fact in the one-tailed test just considered, the null hypothesis should be stated in this directional form: "No subject would perform better under the *A* treatment than under the *B* treatment." We were able to reject the null hypothesis of equal performance with a probability of 1/252. Under the one-tailed null hypothesis just stated, the probability of the *A* treatment having the five best performances could not be greater than the probability under the null hypothesis of equal performance (see pages 137–138 for proof), and so it also can be rejected. Rejection of the one-tailed null hypothesis implies accepting its negation, which is not simply that the treatments do not have the same effect, but that some (one or more) subjects would perform better under the *A* treatment than under the *B* treatment. That is, we conclude that at least one of the subjects given the *A* treatment would have given a lower performance measurement if he had been given the *B* treatment or at least one of the subjects given the *B* treatment would have given a higher performance measurement if he had been given the *A* treatment.

Notice that to generalize to future tests even with the same subjects we must rely on nonstatistical generalization, because we have not taken *the* responses of the subjects now and forever, but simply responses made at a given time.

Example 5.2

To examine in more detail the procedure of testing, let us consider an experiment of the same sort, with only three subjects taking treatment *A* and two taking treatment *B*, where we expect the *A* treatment to be more effective than the *B* treatment. Such sample sizes are artificially small and the test cannot be very sensitive, but the smallness will make it possible to examine the sampling distribution in detail.

We start by assigning a treatment time to each of the subjects. This assignment can be made in some way that makes the treatment times acceptable to both the experimenter and the subjects. Then we put in a box three slips of paper with "A" on them and two slips with "B" on them. After shaking these up, we draw slips of paper from the box to randomly assign a treatment to each person. Then at the time assigned to each person we carry out the experimental treatment randomly assigned to him.

We have predicted that the *A* treatment will be more effective for some subjects than the *B* treatment. Our one-tailed null hypothesis, then, is that for no subject is the *A* treatment more effective than the *B* treatment. That is, no subject assigned to the *A* treatment will get a larger measurement than he would have obtained if he had taken the *B* treatment at that time, and no subject assigned to the *B* treatment would have gotten a larger measurement if he had taken the *A* treatment at that time.

To test the null hypothesis, we will use as a test statistic ΣA (the sum of the *A* scores) to determine the probability, under the null hypothesis, with random assignment of the sort we performed, of getting such a large sum of *A* as we obtain.

Now in fact we cannot specify an exact distribution of equally probable sets of data for our one-tailed hypothesis, since it does not indicate a specific distribution. But pages 137–138 will prove that under our *one-tailed null hypothesis,* the probability of getting ΣA as large as any specified value will be no higher than the probability under this *two-tailed null hypothesis:* Every subject got exactly the same measurement that he would have had if he had been given the other treatment at the time he received his treatment. We can, therefore, work up our distribution of equally probable sets of data under the *two-tailed* null hypothesis and specify that the probability of getting such a large ΣA as the obtained ΣA is no higher under our one-tailed null hypothesis than the probability we get for the two-tailed test.

Our random assignment procedure makes every possible division of our *subjects* into three subjects in A and two subjects in B equally probable. There are 10 ways of dividing five subjects in this manner. Under the two-tailed null hypothesis of identity of treatment effects for every subject, every subject has the same measurement that he would have had under the other treatment; consequently, the null hypothesis in conjunction with random assignment implies that there are 10 equally probable ways of dividing the five *measurements* into three measurements for A and two measurements for B. Suppose the five obtained measurements were 9, 14, and 15 for A, and 8 and 12 for B. The equally probable divisions are the following:

	A	B		A	B		A	B		A	B		A	B
	8			8			8			9			9	
	9	14		9	12		9	12		12	8		12	8
	12	15		14	15		15	14		14	15		15	14
ΣA	29			31			32			35			36	

	A	B		A	B		A	B		A	B		A	B
	8			8			8			9			12	
	12	9		12	9		14	9		14	8		14	8
	14	15		15	14		15	12		15	12		15	9
ΣA	34			35			37			38			41	

For the sample data, $\Sigma A = 38$. Out of 10 equally probable divisions, there are two in which the ΣA is equal to or greater than 38. Therefore, the probability under the two-tailed null hypothesis of getting such a large ΣA as the obtained ΣA is 2/10, and under our one-tailed null hypothesis the probability is *no greater than 2/10*. The result of the experiment would not be significant at the .05 level but would be at the .20 level.

A way to get an intuitive notion of the correctness of counting sums of A equal to the obtained as well as those greater than the obtained ΣA is to regard the obtained ΣA in terms of its "rareness" under the null hypothesis. The probability of getting a ΣA equal to the obtained ΣA or larger is computed because this is a better indication of the "rareness" than the probability of a larger sum. For instance, if the five measurements in groups A and B were all 10s, we could hardly regard an obtained ΣA equal to 30 as rare under the null hypothesis, simply because there were no divisions of the data providing a larger sum; after all, *every* division would provide a ΣA of 30. This distinction between "greater than" and "equal to or greater than" is not necessary for continuous distributions of the kind usually discussed in statistics, but it is an important distinction in dealing with the discrete distributions of outcomes involved in randomization tests. It will be recalled that the same distinction was necessary in specifying the limits of distribution-free confidence intervals.

The preceding reference to "rareness" is not, of course, a *statistical* justification for counting all differences equal to or greater than our obtained ΣA. The statistical justification is that this procedure makes the probability of rejecting the null hypothesis when it is true no greater than the level of significance. For example, in using the .05 level of significance, when the null hypothesis is actually true, the probability is no greater than .05 of rejecting the null hypothesis if we reject it when no more than 5 percent of ΣA are equal to or greater than the obtained sum. On the other hand, the procedure of rejecting every time there are no more than 5 percent of the sampling distribution of ΣA *larger* than the obtained ΣA has a probability *greater than .05* of rejecting the null hypothesis when it is true.

It might be wondered why we used ΣA as our test statistic instead of such other measures as $\Sigma A - \Sigma B$ or $\overline{X}_A - \overline{X}_B$. The reason is that ΣA is easier to compute and will always lead to the same level of significance or probability value as the other test statistics. Even if we decided to use $(\Sigma A)^2$, or the logarithm of ΣA, the test is unaffected. Use of any monotonic (ordinal) transformation of ΣA, linear or nonlinear, as the test statistic will leave the divisions in the same rank order with regard to the magnitude of the test statistic. The division with the largest ΣA has the largest value of $\overline{X}_A - \overline{X}_B$, the largest logarithm of ΣA, the largest value of $\Sigma A - \Sigma B$, and so on. Since the divisions are in the same rank order with regard to the size of the test statistic for all ordinal transformations of the test statistic, and since the probability under the null hypothesis depends only on the rank order of the test statistics over the divisions, the probability is unaffected by any ordinal transformation of the test statistic. This discussion refers to transformations of the *test statistic* only; nonlinear ordinal transformations of *measurements* can affect the probability value or the level at which a result is significant.

Example 5.3

If we were interested in testing the two-tailed null hypothesis that the two treatments differed, rather than a null hypothesis specifying a direction of difference, we would use the procedure commonly followed with two-tailed normal curve tests of using equal-sized rejection regions in the tails of our distribution. We arrange the ΣAs from low to high: 29, 31, 32, 34, 35, 35, 36, 37, 38, 41. For a two-tailed test to be significant at the .20 level, we would have to have either the largest or the smallest ΣA, that is, either 29 or 41. For two-tailed significance at the .40 level, we would have to have one of the two largest sums or one of the two smallest sums, that is, 29, 31, 38, or 41. Our obtained ΣA, 38, is therefore significant at the .40 level, for a two-tailed test.

There is no assurance that there will not be tied sums within the rejection region. For instance, since the distribution of ΣA need not be symmetrical, we could have two 29s for the two lowest scores and a 41 for the highest score. In that case, for a two-tailed test there is no .20 rejection region; .30 is the probability of getting ΣA equal to or less than 29 or equal to or greater than 41. Therefore, it is possible to have a result significant at the .10 level for a one-tailed test, but not significant at the .20 level for a two-tailed test. A procedure for determining two-tailed probabilities in such cases is given in the following example.

Example 5.4

The following procedure can be used with distributions of test statistics for any randomization test, to determine two-tailed probabilities. It is designed to provide equal-sized rejection regions in the two tails of the distribution.

We have 20 equally probable arrangements of measurements, each of which provides a single numerical value of a test statistic. The 20 test-statistic values (for example, the sum of the *A* measurements) are arranged from low to high:

13 14 15 16 16 16 17 18 19 19 20 20 21 23 23 24 24 25 26 26

104 Statistical Inference: The Distribution-free Approach

The first step in the procedure is to draw a single line under the highest value and a single line under the lowest value in the distribution:

13 14 15 16 16 16 17 18 19 19 20 20 21 23 23 24 24 25 26 <u>26</u>

Then extend the line under the 26 so that it passes under both 26s. This is done because there is no reason to regard one 26 as larger than the other. To make the line at the lower end of the distribution the same length, extend it under 14:

13 <u>14</u> 15 16 16 16 17 18 19 19 20 20 21 23 23 24 24 25 <u>26 26</u>

Draw a double line under the third score from each end. Since there is only one 15 and one 25, neither of these double lines needs to be extended:

13 14 <u>15</u> 16 16 16 17 18 19 19 20 20 21 23 23 24 24 <u>25</u> 26 26

Draw a triple line under the fourth score from each end:

13 14 15 <u>16</u> 16 16 17 18 19 19 20 20 21 23 23 24 <u>24</u> 25 26 26

Then extend the triple lines to go under the other 16s and 24s:

13 14 15 <u>16 16 16</u> 17 18 19 19 20 20 21 23 23 <u>24 24</u> 25 26 26

Extend the triple line under the 24s to pass under one more score to make it the same length as the triple line under the 16s:

13 14 15 <u>16 16 16</u> 17 18 19 19 20 20 21 23 <u>23 24 24</u> 25 26 26

Now we see that extending the triple line under the 24s to pass under one more score made it pass under one of a pair of 23s, and so we extend it to pass under both 23s:

13 14 15 <u>16 16 16</u> 17 18 19 19 20 20 21 <u>23 23 24 24</u> 25 26 26

The triple line under the 23s and 24s is now longer than the triple line under the 16s, and so we extend the triple line under the 16s to go under one more score to make the two triple lines of the same length:

13 14 15 <u>16 16 16 17</u> 18 19 19 20 20 21 <u>23 23 24 24</u> 25 26 26

We could continue working inward uniformly from both ends, but the method has been illustrated adequately. Now we use the four scores underlined by the single line as the smallest possible two-tailed rejection region, and so the smallest two-tailed probability for the test-statistic distribution represented is 4/20, or .20. A person using a level of significance of .20 or larger, therefore, could reject the null hypothesis if he obtained a 13, a 14, or a 26. The next larger rejection region consists of the six scores underlined by either a single or a double line. The probability of getting a value in this region is 6/20.

Rejection of the null hypothesis for a two-tailed test only implies that at least one of the individuals would have responded differently under the treatments, and rejection of the null hypothesis for a one-tailed test implies that at least one of the individuals would have responded stronger to the treatment predicted to have the stronger effect. Even though these inferences refer to "at least one of the individuals," not necessarily all or even the majority of them, the inferences may be important, especially in conjunction with follow-up work to determine specifically the kind of individual that differentially responds.

Under what conditions would we expect this test to give significant results? Certainly not in every case where the treatments had a differential effect, even on all individuals, because the difference between the effect of one treatment and the other for an individual could be quite small compared to the differences between individuals under the same treatment. Also, if there were treatment effects in one direction for some individuals and in the opposite direction for other individuals, there would be a tendency for these to cancel out and provide little difference between the totals. Therefore, this test is most appropriate for detecting differences in treatment effects that are large relative to individual differences and in the same direction for the vast majority of the subjects. If only a few subjects are affected by the difference in the treatments and these subjects are affected in the same direction, the test might yield a significant difference, but it would be more likely to do so if several subjects were affected in the same direction. Therefore, we have a more sensitive test when there are a number of individuals with treatment effects in the same direction, with each treatment effect being large relative to the differences between individuals under the same treatment. In other words, our test is most sensitive when the between-treatment variability is large compared to the within-treatment variability.

Randomization test for paired comparisons

Example 5.5

Now let us consider an example with paired groups, instead of independent groups. Suppose we have three individuals and we take a measurement from each under two conditions, *A* and *B*. We could simply test all subjects with condition *A*, and then test all with condition

B, and pair the measurements. We could say that every subject with a difference between the measurements responded differently under A than under B. But this does not answer the question we are interested in. After all, from one treatment administration time to another a subject might give different responses to the *same treatment*, but we are not trying to find out about such variability in responding. Our interest is in the variability produced by treatment differences. Assign two treatment times to each subject. Label two pellets A and B and shake them up in a container. Reach in and without looking select one of the pellets. The pellet selected indicates the treatment to be given a subject on the earlier of the two times. (In this case tossing a coin would be more convenient than using a lottery.) The other treatment is given at the later of the two times. This random assignment is carried out separately for each subject.

First, consider testing the two-tailed null hypothesis that the score of every subject under one treatment is exactly what it would have been under the other treatment if it had been given at that time. Suppose our three subjects provided the following data:

	A	B
Subject 1	12	15
Subject 2	5	7
Subject 3	19	23

Since the treatments were randomly assigned to the two treatment times for each individual, we consider all possible divisions of the six measurements obtained by reversing measurements within subjects. Each division is as probable as any other under the null hypothesis of no treatment effects. There are eight equally probable divisions, shown in the table on the next page. The test statistic we will use to test the null hypothesis is the sum of the differences between the paired A, B scores where A is greater than B, which in the table is "$\Sigma(A > B)$ differences." Our obtained value is 0, which is one of the two most extreme values, namely 0 and 9, and so the probability under the null hypothesis of such an extreme difference between treatments is 2/8, or .25.

We could just as well have used the sum of the differences where *B is greater than A*. Since the sum of the *absolute differences* between paired scores is 9, the sum of the differences where $B > A$ for each of the divisions is simply 9 minus the sum of the $A > B$ differ-

Randomization Tests for Experiments

A	B	A>B difference	A	B	A>B difference	A	B	A>B difference
12	15	—	12	15	—	12	15	—
5	7	—	5	7	—	7	5	2
19	23	—	23	19	4	19	23	—
$\Sigma (A>B)$ differences = 0					4			2

A	B	A>B difference	A	B	A>B difference	A	B	A>B difference
12	15	—	15	12	3	15	12	3
7	5	2	5	7	—	5	7	—
23	19	4	19	23	—	23	19	4
$\Sigma (A>B)$ differences = 6					3			7

A	B	A>B difference	A	B	A>B difference
15	12	3	15	12	3
7	5	2	7	5	2
19	23	—	23	19	4
$\Sigma (A>B)$ differences = 5					9

ences. The rank order of the divisions is just the reverse of the rank order for the size of the sums of differences, and this has no effect on the two-tailed probability value.

The one-tailed probability, of course, depends on which direction of difference is predicted, which determines the null hypothesis to be tested. If our null hypothesis in this study had been that none of the subjects would respond stronger to the *B* treatment than to the *A* treatment, we could reject this hypothesis at the 1/8 (that is, .125) level of significance in favor of the alternative hypothesis that at least one of the subjects would have responded stronger to the *B* treatment than the *A* treatment for at least one of his two treatment times.

Unlike the test for a difference between independent samples, the randomization test for paired comparisons is not affected by individual differences. Just as stratification makes sample estimates more precise by eliminating the effect of variability among strata, paired comparisons provides a more sensitive test for treatment differences than does a test using unpaired measurements when individual differences are great, because the use of pairs eliminates from the test all influence of variability among subjects.

With paired comparisons all subjects receive both treatments, and so no statistical inference can be made about the relative effects of the two treatments in isolation, i.e., if each subject had received only one treatment. If only one treatment were given each subject, as is done in the test for a difference between independent samples, treatments A and B could have the same effect even if the same experimental subjects would have responded differentially to the two treatments in a paired-comparison test. Effects of prior exposure to drugs, material to be memorized, and other "treatments" can make the second of two treatments have a different effect from what it would have had if it had not been preceded by the other experimental treatment. Furthermore, the randomizing of the order of presentation does not necessarily eliminate such sequential effects. Suppose that giving the B treatment first increases the effectiveness of the following A treatment, and giving the A treatment first *does not* increase the effectiveness of the following B treatment. Then use of paired comparisons with random orders for the treatments would tend to make the observed A-treatment effects higher compared to the observed B-treatment effects than would be the case if comparisons were made between two groups of subjects who each received only one of the treatments. The random assignment of a treatment to the earlier treatment time does not eliminate the influence of such sequential effects.

Then why randomly assign the treatments to the treatment times for each subject? The answer is that if there is a good nonstatistical argument to rule out the likelihood of sequential effects, our test permits us to attribute significant differences to inherent differences in treatment effects at a given time. We can conclude from rejection of the null hypothesis that for some subjects, at the time of one or both treatments, the other treatment would not have had the same effect *at that time*. Elimination of sequential effects would eliminate temporal differences as an explanation for the obtained results. On the other hand, suppose that instead of random assignment of treatment times within subjects, we gave one treatment on one day and the other treatment a week later for every subject. Even if we could rule out the likelihood of sequential effects by some empirical argument, there would still be the possibility of other temporal influences on the relative effectiveness of the two treatments, so that the differences between measurements under the two treatments could be solely the result of one treatment being given at a time when the responses would have been stronger to *either treatment*.

Randomization test for contingency

Example 5.6

Qualitative effects of treatments also can be tested with randomization tests. Suppose that we want to compare an experimental and a control treatment to determine whether they have the same effect on mortality. Our "measurement" is determination of whether an animal is alive or dead at the end of one year following the administration of the treatment.

The null hypothesis is that every animal that is alive at the end of a year would have been alive if it had been assigned to the alternative treatment group and every animal that died within the year would have died if it had been assigned to the alternative treatment group. That is, whether an animal lives or dies is independent of the treatment.

We assign a treatment time to each of six animals which have been designated A, B, C, D, E, and F. Then four of the animals are randomly assigned to a control treatment group and the remaining two to an experimental treatment group, for example, a group that received placebos and a group that received inoculations. Suppose that animals B, C, D, and F were assigned to the control group and animals A and E were assigned to the experimental group, and that at the end of a year following the treatment, animals A, C, and E were alive and animals B, D, and F were dead. We can represent the observed results in the following diagram:

	Lived	Died
Control	C	BDF
Experimental	AE	

The random-assignment procedure and the null hypothesis impose the following restrictions on the marginal totals for the diagrams of the equally probable arrangements in the sampling distribution:

	Lived	Died	
Control			4
Experimental			2
	ACE	BDF	

The sampling distribution used to determine the probability, under the null hypothesis, of getting such an extreme result as the obtained will consist, then, of all possible arrangements of the letters A, B, C, D, E, and F within the four cells such that four letters are in the "Control" row, two are in the "Experimental" row, the letters A, C, and E are in the "Lived" column, and the letters B, D, and F are in the "Died" column.

The numbers 4 and 2 for the Control and Experimental totals are fixed by the experimenter's decision to assign four animals to the control group and two to the experimental group. For the rows, only the row *frequencies* are fixed, not the particular animals within each row, because the experimenter randomly assigned the animals to the rows.

The letters A, C, and E for the "Lived" column and B, D, and F for the "Died" column are fixed by the null hypothesis in conjunction with the obtained results. The null hypothesis states that whether an animal lived or died was independent of the group to which it was assigned. Therefore, since animals A, C, and E lived and B, D, and F died in the actual experiment, under the null hypothesis animals A, C, and E would have lived and animals B, D, and F would have died, no matter which groups they were assigned to. Notice that the null hypothesis is not simply that the *number* of animals that live or die is independent of their assignment to groups, but that the independence of mortality and assignment exists for each individual animal. After all, there is no conceivable reason for expecting the number of animals that live and the number that die to remain constant over all possible random assignments unless the *same* animals live or die in each case.

We will now consider the sampling distribution of arrangements from which we will determine the probability of getting such an extreme result as was obtained. There are

$$\frac{6!}{(4!)(2!)} = 15$$

ways the six letters can be divided between the Control and Experimental rows, assigning four to the Control row and two to the Experimental. For each of these 15 assignments, there is only one arrangement of the six letters among the four cells in the diagram because of the restriction specifying which letters must fall in each column. For example, consider the diagram that represents the random assignment of animals A, B, D, and E to the control group and animals C and F to the experimental group:

	Lived	Died
Control	A E	B D
Experimental	C	F

Since letters A, C, and E must fall in the "Lived" column and letters B, D, and F in the "Died" column, the above diagram is the only one in the sampling distribution representing the random assignment of A, B, D, and E to the control group and C and F to the experimental group. The number of equally probable outcomes in the sampling distribution is therefore the number of equally probable random assignments, as is true of all randomization tests.

The fifteen equally probable assignments in the sampling distribution are the following:

(1) ABCD - EF, (2) ABCE - DF, (3) ABCF - DE, (4) ABDE - CF, (5) ABDF - CE, (6) ABEF - CD, (7) ACDE - BF, (8) ACDF - BE, (9) ACEF - BD, (10) ADEF - BC, (11) BCDE - AF, (12) BCDF - AE, (13) BCEF - AD, (14) BDEF - AC, (15) CDEF - AB

We will use as our test statistic the number of control animals that died. There are a number of other ordinally related test statistics that would lead to the same probability values. Examples are the number of experimental animals that died, the number of control animals that lived, and the proportion of control animals that died.

Now, without diagramming, we can see how many of the fifteen arrangements listed earlier have, under the null hypothesis, as many as three control animals that died (the number that died in our experiment). Of the fifteen arrangements these would be arrangements 5, 12, and 14. To test the two-tailed null hypothesis, we also have to consider arrangements that are equally extreme at the other end of the distribution. The three arrangements with the smallest number of dead control animals are arrangements 2, 7, and 9. All together, six of the 15 arrangements are as extreme as the obtained in terms of the number of control animals that died. Thus, if we had set a 40 percent level of significance for a two-tailed test, we could reject the null hypothesis, but not at any stricter level of significance.

If we had results that permitted us to reject the null hypothesis, could we conclude that the inoculation had some effect on the longevity of some of the animals? Not if there are other differences between the treatments than receiving an inoculation or a placebo that could affect mortality. For example, if the animals were kept in separate environments over the year following the treatment administrations, differences in their care and feeding could account for differences in mortality. When we reject the null hypothesis of identical treatment effects, we are considering "treatment" to include everything that goes along with being assigned to that treatment group. It is the responsibility of the experimenter to make sure that his experimental control over extraneous variables permits him to rule out alternative explanations for significant results. Arguments concerning the irrelevance of certain conditions associated with one treatment and not associated to the same degree with the other must of course be nonstatistical.

Example 5.7

Now consider a test of a one-tailed null hypothesis.

Suppose we believe that for at least one of the animals, assignment to the experimental treatment would result in the animal's being alive, whereas it would have been dead if it had been assigned to the control group. Then we specify this null hypothesis: For every animal, either (1) the two treatments would have the same effect on its mortality or (2) it would have been alive if it were assigned to the control group but dead if it were assigned to the experimental group.

We can use as our test statistic the number of dead control animals, which we expect to be relatively large. In the sampling distribution provided by the *two-tailed* null hypothesis, the probability is 3/15 or .20 of getting as many as three dead control animals, the number observed in the experiment. Within the sampling distribution for the one-tailed null hypothesis the frequency in the Control-Died cell in each of the 15 arrangements will be *equal to or less than* the frequency for the corresponding arrangement under the two-tailed null hypothesis. Thus, the probability of as many as three dead control animals under the *one-tailed* null hypothesis is *no greater than* .20.

Example 5.8

In the randomization test for contingency, we used a sampling distribution consisting of every possible arrangement of individuals within the following row and column restrictions:

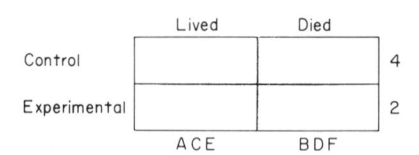

	Lived	Died	
Control			4
Experimental			2
	A C E	B D F	

These restrictions limit the sampling distribution to those arrangements that our random assignment procedure and null hypothesis make equally probable.

Suppose that we had used a sampling distribution based on the following row and column restrictions:

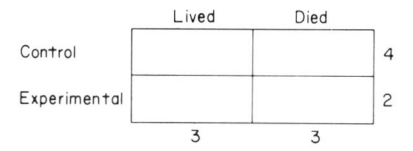

	Lived	Died	
Control			4
Experimental			2
	3	3	

Such a sampling distribution would contain not only those arrangements where A, C, and E are in the "Lived" column and B, D, and F are in the "Died" column, but arrangements for every way of dividing the six letters so that three are in each column. For every one of the 15 equally probable treatment assignments there would be 6!/3!3! = 20 ways of dividing the letters between the two columns, making a total of 300 arrangements in the sampling distribution which are regarded as equally probable. Even within the random assignment actually performed in the experiment there would be 20 different sets of three letters in the "Lived" column that are to be regarded as equally probable under the null hypothesis. The null hypothesis that would be tested would then seem to be this: For every possible way the individuals could be assigned to the two treatments, three individuals would be alive at the end of one year, but not necessarily A, C, and E—all possible sets of three individuals are equally probable to be the individuals that survive. Clearly this null hypothesis differs from our deterministic null hypothesis which specified for each animal that whether it lived or died was unaffected by the treatment to which it was assigned. We deliberately introduced a random process into our experiment in the form of a lottery, and that is the basis for regarding the 15 assignments as equally likely. But the basis for regarding 20 different combinations of individuals as equally likely to die for a particular random assignment is not the introduction of a man-made randomizing device. The basis seems to be the assumption of a natural probabilistic process.

There is, in fact, a frequently used statistical test that corresponds to the hypothetical test we have been discussing. The test is Fisher's exact test, which involves fixed row and column totals but in no way specifies the individuals that must fall within a designated column.

The previous discussion shows that Fisher's exact test must have a probabilistic null hypothesis in contrast to the deterministic null hypothesis for the randomization test for contingency. We will now see what effect the difference between marginal restrictions and the consequent difference between the sampling distributions for Fisher's exact

114 Statistical Inference: The Distribution-free Approach

test and the randomization test for contingency has on the determination of the probability of a particular cell frequency. For this comparison we will use the situation described in Example 5.6 and will determine the probability, under the null hypothesis, of exactly three dead control animals, in other words, the probability of a frequency of 3 in the upper right cell.

In Example 5.6 we had 15 arrangements where *B, D,* and *F* (the animals actually dead at the end of a year) were always in the right ("Died") column. These 15 arrangements will be used to determine the probability provided by the randomization test for contingency of three individuals in the upper right cell. Examination of the list of 15 configurations shows that in three of them all three animals *B, D,* and *F* in the right column are in the upper cell. Thus the probability of three dead control animals is 3/15. That is also the probability of *as many as* three dead control animals inasmuch as there are no more than three dead control animals in any arrangement in the sampling distribution.

Suppose we repeated our randomization test for contingency (*not* the experiment) 20 times, with the same data, for the 20 different ways of dividing letters *A, B, C, D, E,* and *F* between the two *columns* so there are three letters in each column. This need not be regarded as a reclassification of animals with regard to the "Lived" and "Died" categories. It may be regarded as nothing more than a *relabeling* of our animals. Our test is independent of the assignment of labels (that is, letters) to the animals, and so we would then have 20 sets of sampling distributions with identical *frequencies* within the cells. (The same animals are represented in the same cells for a particular random assignment in each of the sets, but their individual labels vary from set to set.) Now, these 20 sets constitute the 300 arrangements of Fisher's exact test. The identity of these 20 sets (one of which is the set of 15 arrangements we used earlier in our test) with regard to cell frequencies implies that within each set there will be the same number of arrangements with three dead control animals. These will be the arrangements corresponding to the ones in our set which contain three dead control animals. Since 3/15 of the 15 arrangements in our set have three dead control animals, then within each of the other 19 sets 3/15 of the 15 arrangements must have three dead control animals. Thus 3/15 of the total 300 arrangements must have three dead control animals.

Thus the probability of getting exactly three dead control animals is 3/15 for both the randomization test for contingency and for Fisher's exact test. And the same sort of reasoning shows that the two tests will always give the same probability for any specified cell frequency. To get the probability for a cell frequency *as large as* an obtained frequency in order to determine its significance, one adds the probabilities for all cell frequencies as large as the obtained frequency. Since each of these probabilities is the same for Fisher's exact test and the randomization test for contingency, the sums must be the same for the two tests. Thus the two tests lead to identical significance values.

The equivalence of the probabilities determined by Fisher's exact test and by the randomization test for contingency permits the use of computational formulas for Fisher's exact test in testing the null hypothesis specified by the randomization test for contingency. For example, we can use the following formula of Fisher's to determine the probability of a particular configuration of cell frequencies:

$$\frac{(a+b)!\,(c+d)!\,(a+c)!\,(b+d)!}{N!\,a!\,b!\,c!\,d!}$$

where the symbols refer to the following frequencies within the 2 × 2 table:

a	b	(a+b)
c	d	(c+d)
(a+c)	(b+d)	N

N is the number of individuals in the table and a, b, c, and d in Fisher's formula refer to the frequencies in the corresponding cells of the table. For example, to use this formula to find the probability of three dead control animals, we would compute

$$\frac{4!\,2!\,3!\,3!}{6!\,1!\,3!\,2!\,0!}$$

to get 3/15 as the probability.

Thus the formulas for Fisher's exact test can be used to test the deterministic null hypothesis of the randomization test for contingency: the observed outcome for each individual is what it would have been under assignment to the other treatment.

Randomization test for correlation

Randomization tests can also be applied to correlational problems of an experimental nature. For example, we may want to find out if variation in the magnitude of an independent variable produces variation in a dependent variable. We first assign each subject a treatment time. Then we randomly assign each subject to a magnitude of the independent variable. After taking measurements of the dependent variable, we pair them with the magnitudes of the independent variable. Then we compute a correlation statistic to use as our test statistic. The value of this correlation statistic is compared to the values under the null hypothesis which are obtained by pairing every stimulus magnitude with every response magnitude and calculating the correlation statistic.

The *two-tailed null hypothesis* is that there is no effect of variation of the stimulus variable on the response; that is, for every subject the response given is the same as it would have been if he had been given any of the other magnitudes of the stimulus at his treatment time. Rejection of this null hypothesis implies acceptance of the alternative hypothesis that at least one of the subjects would have given a different response to at least one of the other magnitudes of the stimulus.

The *one-tailed null hypothesis* to be used when a positive correlation is expected is this: No subject would have given a larger response measurement if he had been assigned to a larger stimulus magnitude for his treatment time, or given a smaller response measurement if he had been assigned to a smaller stimulus magnitude at that time. Rejection of the one-tailed null hypothesis implies accepting the alternative hypothesis that at least one of the subjects would have given a larger response measurement if he had been assigned to a larger stimulus magnitude at his treatment time or would have given a smaller response measurement if he had been assigned to a smaller stimulus magnitude at that time.

The two-tailed null hypothesis for this test and those for the other randomization tests in this chapter have been expressed in terms of *identity* of treatment effects. Such a null hypothesis is not as untenable as it first appears. The identity of treatment effects refers to a particular time at which an individual is given a treatment. Consequently, this null hypothesis is consistent with the assumptions of variability within individuals over time and variability from individual to individual.

The *one-tailed* null hypothesis for correlation will sometimes be plausible when the two-tailed null hypothesis is not, and so it is likely to be used more frequently than the two-tailed null hypothesis. The same is true of null hypotheses for other randomization tests.

If a person is fairly certain that the treatments under comparison will have at least slightly different effects and is also sure of the direction of the difference, he has no reason to use a hypothesis-testing procedure. He should use, instead, a technique for determining confidence intervals on the amount of difference between the treatments or some other technique more relevant to his purpose than hypothesis testing.

Randomization Tests for Experiments 117

Example 5.9

We assign each of four subjects a time to be tested. Then we randomly assign four stimulus magnitudes, 5, 10, 15, and 20, to the subjects, which are to be presented at the times already assigned to the subjects. The following table shows the obtained data:

Subject	Stimulus magnitude	Response magnitude	Product of stimulus and response magnitudes
A	5	15.6	78.0
B	10	13.8	138.0
C	15	20.4	306.0
D	20	31.5	630.0
			1,152.0

We can use the sum of the product of the stimulus and response magnitudes ("sum of the cross products") as a correlation statistic because in pairing numbers, the maximum sum of cross products occurs when the numbers are arranged in the same rank order and the minimum sum when they are in opposite order.

We work up our distribution of sums of cross products under the two-tailed null hypothesis by considering every equally probable pairing of subjects and stimulus magnitude. Now the null hypothesis implies that a subject's response is the same as it would be for any other stimulus magnitude, and so every equally probable pairing of subjects and stimulus magnitude is to be represented by every pairing of the stimulus values with the obtained response measurements. Thus, each of the four stimulus magnitudes can be paired with each of the four response magnitudes, making 4! = 24 different equally probable pairings. For each of these pairings we compute the cross product of the stimulus and response magnitudes and the sum of these cross products. Of the 24 pairings, only 2 would give a sum of cross products as large as 1,152: the pairing in our sample and the one where the response magnitudes are in the same rank order as the stimulus magnitudes. For the two-tailed test, we consider also the two most extreme sums of cross products in the other tail of the distribution, namely, the two smallest sums of cross products. The smallest sum of cross products is 871.5, resulting from the pairing of the stimulus and response magnitudes in inverse order. The next larger sum would be obtained when the stimulus and response magnitudes were paired in inverse order, except for 15.6 being paired with 20 and 13.8 being paired with 15. There are four sums of cross products as extreme as the sum that was obtained, and so the

118 Statistical Inference: The Distribution-free Approach

probability under the two-tailed null hypothesis of such an extreme sum of cross products is 4/24.

If we predicted a positive correlation, we should determine the probability under the previously stated one-tailed null hypothesis of getting as large a cross product as 1,152. Since there are just two sums of cross products as large as 1,152, the probability is 2/24 of getting such a large sum of cross products under the *two-tailed* null hypothesis. The *one-tailed* null hypothesis can have no higher probability of getting such a large sum of cross products, and so the probability under the one-tailed null hypothesis is *no greater than 2/24*. The null hypothesis could be rejected at the .10 level, for example, but not the .05 level, for a one-tailed test.

Randomization test for interaction
Example 5.10

Under certain conditions a randomization test for interaction can be performed. For instance, assume that we want to find out whether the differential effect of A and B is the same as that of C and D. Randomization tests cannot test null hypotheses about mean effects (although a difference between sample means may be used as a *test statistic*), and so instead of the null hypothesis $\mu_B - \mu_A = \mu_D - \mu_C$, our two-tailed null hypothesis is that for *every individual*, $B - A = D - C$. To be more specific, our null hypothesis is that each individual given A treatment and B treatment has a differential score $B - A$ equal to the differential score $D - C$ he would have had if he had been given C and D instead, and each individual given C and D has a differential score equal to what he would have had with A and B. We assign each of 10 subjects two treatment times, the earlier to be used for treatment A and the later for treatment B if he is subsequently assigned to the A, B group, or the earlier for treatment C and the later for treatment D if he is subsequently assigned to the C, D group. Then we randomly assign five subjects to the A, B group and the remaining five to the C, D group. Suppose we obtained the following scores:

Subject	A	B	B−A	Subject	C	D	D−C
A	5	8	3	F	10	16	6
B	4	7	3	G	9	13	4
C	9	9	0	H	14	12	−2
D	6	5	−1	I	15	19	4
E	7	10	3	J	17	22	5
			$\Sigma(B-A)=8$				$\Sigma(D-C)=17$

The obtained $\Sigma(B-A)$ is 8, while the obtained $\Sigma(D-C)$ is 17, slightly more than twice as large. When there is an interaction in which subjects under one of the two conditions (A, B and C, D) show larger differential scores in general than under the other, a large discrepancy between $\Sigma(B-A)$ and $\Sigma(D-C)$ is expected. Therefore, we can use the absolute difference between $\Sigma(B-A)$ and $\Sigma(D-C)$ as our test statistic.

To test the null hypothesis we consider every possible division of the 10 subjects between the A, B and C, D groups, assigning five subjects to each group. There are 10!/5! 5!, or 252 equally probable divisions. Now the null hypothesis *does not* imply that a subject's obtained *scores* are the same as they would have been if he had been assigned to the other group. For example, it is unnecessary to assume that subject B would have gotten a 4 under treatment C and a 7 under treatment D, if he had been assigned to the C, D group. The null hypothesis does imply, however, that subject B would have a difference $D-C$ equal to 3, which is subject B's difference score for $B-A$. The null hypothesis implies that the *differential scores* in the last column are the same for each subject as they would have been if he had been assigned to the other group. Thus, under the null hypothesis, there are 252 equally probable divisions of the 10 differential scores in the last column, since each *differential score* is assumed to go with the subject to whichever group he is assigned. We determine what proportion of the 252 divisions provide a difference between $\Sigma(B-A)$ and $\Sigma(D-C)$ as large as 9, which is what we obtained. If this proportion is no greater than 5 percent, we can reject our two-tailed hypothesis at the .05 level, in favor of the alternative hypothesis that for at least one of the 10 subjects, the differential score for A, B would be different from the differential score for C, D for the times at which he received the experimental treatments.

This example of a randomization test for interaction was given simply to illustrate a special application of a randomization test. The distribution-free approach to statistical inference, however, must take into consideration the purpose of making statistical inferences, and there seems to be no good reason why one would want to test the null hypothesis specified above. If we expected for example that C and D treatments provided larger measurements than A and B treatments, it is hard to imagine why we would think it possible that the differential effect of the C and D treatments would be the same as the differential effect of the A and B treatments for every subject. And why carry out a test to reject an extremely implausible null hypothesis?

To reduce the possibility of misunderstanding, it must be remarked that the implausibility of the null hypothesis for this randomization test is no greater than the implausibility of the null hypothesis for the normal curve test of significance of interaction. When the mean effects of D and C are expected to differ from the mean effects of A and B, there is no basis for expecting $\mu_B - \mu_A$ to be equal to $\mu_D - \mu_C$.

But a *one-tailed* null hypothesis stating that $B - A$ is equal to *or greater than* $D - C$ for every individual in the experiment, and the corresponding normal curve null hypothesis about means, might be worth testing.

Other randomization tests

The randomization test for interaction was concerned with a null hypothesis about differential effects, but randomization tests for experiments would usually involve null hypotheses about differences between two single treatment effects. For this purpose one can use a number of tests in addition to those already discussed. For example, a person could construct a randomization test for which the test statistic is a difference between medians, a difference between ranges, or a difference between standard deviations. Even though the test statistic varies from one situation to another, the null hypothesis is always identity of treatment effects for all individuals, and rejection implies that for some subjects the treatments would not be equally effective. Differences between means, between medians, and between ranges are used as test statistics for testing hypotheses about *individuals, not* about means, medians, and ranges. One test statistic is more sensitive than another for certain types of differences in treatment effects, but the null hypothesis and alternative hypothesis do not refer to the test statistic at all.

We cannot test the null hypothesis that the mean measurement under treatment A is the same as the mean measurement under treatment B for a particular group of individuals by means of a randomization test. Randomization tests are distribution-free, and the sampling distribution of differences between means is affected not only by the means of the treatments, but also by the variability, skewness, and other characteristics of the distribution that are not specified by a null hypothesis of identity of means. The following example will show how the sampling distribution of differences between means is affected by differences in shapes of distributions of measurements for two treatments.

Example 5.11

Suppose we have the following two *hypothetical* distributions of measurements under treatments A and B for eight subjects. The two distributions *do not* represent two sets of *obtained* scores. They represent for each subject two measurements that *could be obtained*, one measurement if he were given treatment A at his treatment time and the other the measurement that he would have if he were given treatment B at his treatment time instead of treatment A. The distributions have *identical means*.

Subject	A	B
a	25	20
b	25	20
c	25	20
d	25	20
e	25	20
f	25	20
g	25	40
h	25	40
$\overline{X}$ =	25	25

Now let us consider the sampling distribution of $\overline{X}_A - \overline{X}_B$ for all equally probable divisions of the eight subjects into six for treatment A and two for treatment B. There are 8!/6! 2!, or 28, different equally probable divisions. How often would we get $\overline{X}_A - \overline{X}_B$ as large as 5? Since all subjects assigned to A would get 25, and since no subject assigned to B would get less than 20, we would get a difference as large as 5 between the two means every time we got two 20s in the B treatment. The probability of this is 15/28, and so the probability of getting a difference between means of as much as five points in favor of A treatment is 15/28, or approximately .54.

Now suppose with our random assignment we obtained the following set of measurements, which we have seen to have a probability of .54:

	A	B
	25	20
	25	20
	25	
	25	
	25	
	25	
$\overline{X}$ =	25	20
$\overline{X}_A - \overline{X}_B$ =	5	

122 Statistical Inference: The Distribution-free Approach

What probability would we get by using the randomization test for a difference between independent samples, for such a large $\overline{X}_A - \overline{X}_B$ as the obtained value of 5? In testing this we would consider the 28 equally probable divisions of *these 8 numbers* into two groups with 6 for A and 2 for B to see how often we would get such a large $\overline{X}_A - \overline{X}_B$. Such a large difference would occur only for the obtained division, in which both 20s are in the B group. So the probability is said to be 1/28 (or about .04) of getting a $\overline{X}_A - \overline{X}_B$ as large as 5 under the null hypothesis.

But we have already seen that the probability, given the situation we describe, is .54, considerably higher than .04. This implies that if someone were to carry out an experiment where the hypothetical distributions are like those in the first distribution, he has a probability of .54 of getting a result that he will consider to be significant at the .04 level. What is wrong? Nothing is wrong; the randomization test would never *falsely* reject the null hypothesis because the null hypothesis of identical measurements under the two treatments *is* false—the means are identical, but the measurements for individuals are not identical. The two hypothetical distributions for treatments A and B under the randomization-test null hypothesis would be identical.

On the other hand, consider the person who uses the randomization test, using $\overline{X}_A - \overline{X}_B$ as his test statistic, and decides to reject the null hypothesis of *identity of means,* at the .04 level. The treatment means are identical, and so his null hypothesis is *true;* yet, the person has a probability of .54 of rejecting it at the .04 level. The use of randomization tests for rejecting null hypotheses about *means,* then, can lead to spuriously high significance levels.

When we use a randomization test for a difference between independent samples, the probability of rejecting the null hypothesis at the .05 level may be considerably larger than .05 when the treatment means are identical, if the distributions differ in other respects. For this reason, we cannot imply from rejection of a randomization-test null hypothesis that the treatment means differ, or that the treatment ranges differ, or that some other statistic differs. We simply conclude that the treatments are not identical for all subjects.

Which would be a better test for detecting a difference in variability: subtracting the standard deviation of one group from that of the other or dividing one standard deviation by the other? One test statistic would be as sensitive as the other, but the ratio of standard deviations seems more appropriate since it is independent of the unit of measurement, any given ratio being the same, for instance, whether you use an inch or a foot as your measuring unit. One might think that a ratio of variances would be even better, but the probability values are exactly the same as with the ratio of standard deviations, since the ratio of variances is simply the square of the ratio of standard deviations, and squaring leaves all ratios in the sampling distribution of equally probable ratios in the same rank order.

Rank-order tests

We will now deal with the topic of rank-order tests, a topic which most readers will find more familiar than randomization tests. Rank-order tests, however, are simply special randomization tests—randomization tests using ranks of measurements instead of the original measurements. Rejection of the null hypothesis of identity of ranks under two treatments implies that at least two of the individuals would have had a different rank under the other treatment. We refer to two individuals rather than one, because the rank of one individual cannot change without changing that of at least one other individual. Now the relative positions of two individuals cannot change unless there is some effect of the difference between treatments, and so rejection of the null hypothesis for a rank-order test implies that at least one individual would have had a different measurement if he had been given a different treatment at his treatment time.

The advantage of a rank-order test over a randomization test using the original measurements lies in the ease of determining the probability values. It has been practical to construct probability tables for rank-order tests because the same set of rank numbers is used over and over. For example, the same probability table can be used to test for a difference between two groups of five measurements each, whatever the value of the actual scores, since in every case the sampling distribution of the rank-order statistic is based on random partitions of the rank numbers 1 to 10 into two groups of five ranks each.

Rank-order test for independent samples

In the randomization test for independent samples, we used the sum of the *A* measurements as our test statistic. We could perform the same test if all of the measurements for the *A* and *B* treatments were combined and ranked, giving a rank of 1 to the *smallest* measurement, 2 to the next larger, and so on. Then we would make all possible divisions of the set of ranks and compute the sum of the *A ranks*. The test would then be equivalent to the Mann-Whitney *U* test, the computed probability

124 Statistical Inference: The Distribution-free Approach

value being exactly the same. This is because rank ordering all possible divisions of the measurements with regard to the size of the U statistic also rank orders the divisions with regard to the size of the sum of A ranks. In some books (for example, Meredith, 1967) tables for the Mann-Whitney U test contain probabilities for sums of ranks.

Consider the steps in the Mann-Whitney U test. First we combine, say, two samples of five measurements each, into a joint distribution of ten measurements. Then we rank the measurements from low to high, giving the rank of 1 to the smallest measurement, 2 to the next larger, and so on. Then we discard the original measurement numbers and use only the ranks. We compute the sum of ranks that is appropriate for our one-tailed or two-tailed test or we compute the U statistic and refer to statistical tables to determine the probability of getting such an extreme value under the null hypothesis, for two samples of five measurements each. The probability from the table is the same value that we would get if we performed the randomization test for independent samples on the *ranks of our measurements,* but not necessarily the same as we would get when applying the randomization test to the *original measurements.*

Rank-order test for paired comparisions

Example 5.12

The rank-order counterpart of the randomization test for paired comparisions is the *Wilcoxon matched-pairs signed-ranks test.* Instead of rank ordering the original measurement numbers, we use the original measurement numbers to determine a distribution of $A - B$ differences. Then we rank order all $A - B$ differences according to their absolute values. *For this test the ranks must be assigned from low to high;* that is, the *smallest* difference gets a rank of 1 and the *largest* a rank of N. The test statistic is the sum of the ranks of the $A - B$ differences where A is greater than B. For instance, suppose that we have the following data from treatments A and B:

Subject No.	A	B	A − B	|A − B|ranks
1	29	20	9	2
2	15	28	−13	(4)
3	18	16	2	1
4	35	23	12	3
			Sum of ranks where $A > B = \overline{6}$	

Reference to probability tables can be made to determine the probability of getting as large a sum of ranks for $A > B$ as 6, for paired comparisons with four subjects.

The above statistic, the sum of the ranks where $A > B$, is called Wilcoxon's T. The probability of getting a T as large as 6 under the null hypothesis of no treatment difference is the proportion of sums of ranks where $A > B$ which are equal to or greater than 6, in the sampling distribution obtained by this procedure: For ranks 1 to 4, assign a plus or a minus to each rank in every possible way. There are 2^4, that is, 16, different equally probable arrangements under the null hypothesis:

	+1	+1	+1	+1	+1	+1	+1	+1
	+2	+2	+2	+2	−2	−2	−2	−2
	+3	+3	−3	−3	+3	+3	−3	−3
	+4	−4	+4	−4	+4	−4	+4	−4
Sum of positive ranks =	10	6	7	3	8	4	5	1

	−1	−1	−1	−1	−1	−1	−1	−1
	+2	+2	+2	+2	−2	−2	−2	−2
	+3	+3	−3	−3	+3	+3	−3	−3
	+4	−4	+4	−4	+4	−4	+4	−4
Sum of positive ranks =	9	5	6	2	7	3	4	0

Seven out of the sixteen equally probable arrangements contain a sum of positive ranks (representing cases where $A > B$) as large as 6, and so the probability is 7/16. Now, if we had obtained a sum of 10, we would have a probability of 1/16. In paired comparisons the number of arrangements is 2^N, where N is the number of pairs, and so a one-tailed probability in paired comparisons can never be less than $1/2^N$, whether it is concerned with original measurements or ranks. The tables of Wilcoxon's T are derived by the above procedure for various numbers of pairs.

Contingency chi-square test

The randomization test for contingency dealt with categorical data, and so there is, of course, no rank-order counterpart. The *contingency chi-square test* however is frequently used for a similar purpose. The advantage of the contingency chi-square test over the randomization test for contingency is the ready availability of chi-square probability tables giving a close approximation to the exact probability of getting a

126 Statistical Inference: The Distribution-free Approach

contingency chi-square statistic as large as some specified value, under the null hypothesis of no difference in treatment effects.

Example 5.13

Consider the data from Example 5.6:

	Lived	Died
Control	C	B D F
Experimental	A E	

We determined that the probability of getting as extreme a number of dead control animals as three, under the *two-tailed* null hypothesis, was .40. The contingency chi-square test uses chi-square as a test statistic, and consequently is a two-tailed test. We will now compute a contingency chi-square probability on the same data to compare with the two-tailed probability of .40. First we express the observed data in Example 5.6 as frequencies within cells:

	Lived	Died
Control	1	3
Experimental	2	0

In computing the contingency chi-square statistic, we first construct a table similar to this one, showing not the observed frequencies but the frequencies in each of the four cells that would be expected if the treatments had no effect on whether an animal lived or died. The *expected frequencies* are unbiased estimates of cell frequencies under the null hypothesis, for our row and column total frequencies. If the treatments have no differential effect on mortality, we would expect that if half of all of the animals were to die, half of the control animals would die and half of the experimental animals would die. Similarly, we would expect half of the control animals to live and half of the experimental animals to live. (We say half because the column totals show that in our study, 1/2 of all the animals died.) The following table shows the expected frequencies:

	Lived	Died
Control	2	2
Experimental	1	1

If we were to average the Control-Died cell frequency over all 15 equally probable arrangements given in Example 5.6, we would find that the average is 2. Similarly for the Control-Lived cell. The Experimental-Lived cell averages out to 1 and the Experimental-Died cell averages out to 1 also, over the 15 arrangements. This is the technical meaning of these frequencies being "expected frequencies."

The contingency chi-square value is based on the absolute difference between corresponding cells in the table of observed frequencies and the table of expected frequencies. Each absolute difference is reduced by .5 and the resulting value is squared. Then the number resulting from the squaring is divided by the expected frequency for that cell. This procedure is followed for all four cells and the final results are added together to give a chi-square value. In this example, the contingency chi-square value is:

$$\frac{(|1\text{-}2| - .5)^2}{2} + \frac{(|3\text{-}2| - .5)^2}{2} + \frac{(|2\text{-}1| - .5)^2}{1} + \frac{(|0\text{-}1| - .5)^2}{1} = .75$$

The significance of the contingency chi-square value is determined by referring to the row for one degree of freedom in a table of chi-square. The chi-square value of .75 is barely above the value of .708 required for significance at the .40 level. This is consistent with our exact two-tailed probability of .40.

The tabled chi-square probabilities are approximations to the exact probability of getting a contingency chi-square value as large as some value. The exact probability for a 2 × 2 table can be computed by taking every equally probable arrangement of the individuals within columns, keeping the marginal totals fixed, and computing a contingency chi-square statistic for each arrangement. The distribution of all equally probable contingency chi-square statistics can then be used to determine the significance of the obtained value.

For the data in Example 5.6 the contingency chi-square computed over all 15 arrangements would show exactly 6 arrangements that would give a contingency chi-square value as large as .75—the same arrangements used in that example to determine the two-tailed probability of getting as extreme a number of dead control animals, namely, the arrangements with either 3 dead control animals or 1 dead control animal. However, the two randomization test procedures, one using the frequency in cell b as the test statistic and the other using contingency

128 Statistical Inference: The Distribution-free Approach

chi-square as the test statistic, will not always give the same results. The size of the contingency chi-square statistic depends on deviations from the expected frequencies, and so arrangements that deviate in the same degree from the table of expected values will have the same value of chi-square. On the other hand, the arrangements that gave the same contingency chi-square value might not be judged equally extreme by the procedure in Example 5.4 for constructing two-tailed rejection regions, which is intended to make the two rejection regions equal in size, rather than equidistant from the expected values.

Rank correlation

Rank-correlation methods can be used as substitutes for randomization tests for correlation. There is no rank-correlation procedure which uses the sum of cross products of ranks as its test statistic, but such a procedure would be the direct counterpart of the randomization test for correlation described in Example 5.9. Such a rank-order correlation technique would require pairing the scores, assigning ranks within each set of scores separately, and then multiplying the paired ranks. The sum of the cross product of the paired ranks would be the correlation test statistic used to determine the significance of an obtained sum of cross products of ranks.

Example 5.14

Suppose we used three subjects in an experiment in which we obtained one response measurement from each subject and each subject was given a different stimulus magnitude:

Subject	Stimulus	Response	Stimulus Ranks	Response Ranks	Cross products of S and R Ranks
a	10	17	3	3	9
b	20	19	2	2	4
c	30	24	1	1	1
					14

Under the null hypothesis that the response is not affected at all by the stimulus magnitude, and with random assignment of subjects to stimulus values, each pairing of stimulus and response magnitudes is equally probable, and each pairing of stimulus and response ranks is equally probable. There are six equally probable pairings of ranks:

S	R	S×R	S	R	S×R	S	R	S×R	S	R	S×R	S	R	S×R	S	R	S×R
1	1	1	1	1	1	1	2	2	1	2	2	1	3	3	1	3	3
2	2	4	2	3	6	2	1	2	2	3	6	2	1	2	2	2	4
3	3	9	3	2	6	3	3	9	3	1	3	3	2	6	3	1	3
		14			13			13			11			11			10

Only one of the six equally probable sums of cross products of ranks is as large as 14, the obtained sum, and so the probability under the null hypothesis of so large a sum of cross products of ranks is 1/6. Obviously the above distribution of sums of cross products of ranks could be used to determine the significance of a sum of cross products of ranks for any data involving three pairs of scores. Similar distributions for four pairs of scores, five pairs, and so on, would permit the construction of a probability table to be used with such a rank-correlation technique.

It is unnecessary, however, to construct a probability table for the sum of cross products of ranks, because for any specified number of pairs of measurements the probability value associated with a sum of cross products of ranks is exactly the same as that associated with the rank-correlation coefficient known as Spearman's rho. The formula for Spearman's rho is

$$\rho = 1 - \frac{6 \Sigma D^2}{N(N^2 - 1)}$$

where the ΣD^2 is the sum of the squared differences between paired ranks, and N is the number of pairs. Notice that for any given N (number of pairs), the Spearman rank-correlation coefficient varies inversely with the size of ΣD^2—*inversely*, because the fraction of which 6 ΣD^2 is the numerator is subtracted from 1. As ΣD^2 increases, the value of rho *decreases*. Therefore, if we were interested only in the probability of getting a given degree of relationship between two sets of ranks and not in a correlation coefficient as such, we could use ΣD^2 as a correlation statistic. The larger the size of ΣD^2, the *less* the degree of positive correlation, and so the strength of positive relationship would be indicated by the *smallness* of ΣD^2.

Now let us see the relationship between ΣD^2 and ΣXY as correlation statistics:

and so
$$\Sigma D^2 = \Sigma(X - Y)^2 = \Sigma(X^2 - 2XY + Y^2)$$
$$\Sigma(X^2 - 2XY + Y^2) = \Sigma X^2 - 2\Sigma XY + \Sigma Y^2$$
$$\Sigma D^2 = \Sigma X^2 - 2\Sigma XY + \Sigma Y^2$$

Now, for any given number of pairs of ranks, over all possible pairings, ΣX^2 and ΣY^2 remain constant because each is simply the sum of the squares of ranks 1 to N. Thus we see that the term $-2\Sigma XY$ is the only one that varies over the sampling distribution of all pairings. ΣD^2 therefore varies *inversely* with ΣXY, and so the probability of getting a ΣXY as large as some obtained value is exactly the same as the probability of getting as large a value of Spearman's rho as the obtained rho.

Tied ranks

Probability tables for rank-order statistical tests presume that each measurement in a set that is ranked has a different rank from every other one in that set. For example, for rank correlation, there are two sets of measurements which are ranked separately, and the probability tables are based on the assumption that the members of one set of measurements have the ranks 1 to N and that the members of the other set of measurements also have the ranks 1 to N where N is the number of pairs of measurements. This implies that every rank number from 1 to N is used within each set, giving every measurement within a set a different rank number from every other one. For the Mann-Whitney U test for differences, on the other hand, there is just one set of ranks assigned. All measurements in both groups are combined and ranks 1 to N are assigned, where N is the total number of measurements in the two groups. This implies that every measurement has a different rank from every other measurement in either of the groups.

It is not always possible to assign different ranks, though, because two measurement values may be identical. There are several procedures that are followed in dealing with such situations, which will be examined separately.

Now ties *within groups* have no effect on the probability for the U test. For example, suppose that after the first five ranks have been assigned in the joint distribution of A and B measurements, there are three group A measurements tied for the next rank. They can be assigned ranks 6, 7, and 8 in an arbitrary manner because the sum of ranks for group A, and consequently the probability, are unaffected. Settling ties *between groups A and B* in this manner would of course be inappropriate because the probability is affected by which group receives the higher ranks when the ties are broken.

One procedure that is sometimes followed when it is impossible to arrange all measurements in rank order because of tied measurements is to discard the tied measurements, and to carry out the test with a reduced N. There is no reason to believe that this procedure introduces bias, but researchers are reluctant to throw away data unless they have to.

An alternative procedure is to use a random process for assigning ranks to the tied measurements that will provide the measurements with the proper-sized ranks relative to those with unambiguous ranks and will at the same time not affect the probability of rejecting the null hypothesis when it is true.

Example 5.15

For example, in the case of computing a Spearman rank correlation, suppose we have the following data:

X	Y	X rank	Y rank
46	25	1	1
34	22	2	2
23	16	3 or 4	3
23	13	3 or 4	4

We will assign one of the two 23s a 3 and the other a 4, because there are two X measurements that are higher in rank. We toss a coin to determine which is to be assigned rank 3 and which rank 4. The following two outcomes are equally probable:

X rank	Y rank	X rank	Y rank
1	1	1	1
2	2	2	2
4	3	3	3
3	4	4	4

The first outcome has a Spearman rank correlation of .80. The probability of such a large positive correlation under the null hypothesis is 3/24. The second outcome has a Spearman rank correlation of 1.00 and the probability of such a large positive correlation is 1/24. Now this random assignment procedure provides ranks that might have been obtained if our measuring instruments had been able to make a distinction between the two measurements designated as 23. One can argue that if one of the measurements designated as 23 should have been larger than the other, it is just as probable that it is the one paired with 16 as the one paired with 13. This way of settling the problem of ties does not affect the probability value at all. That is, when the null hypothesis of no stimulus effect is true and when there is random assignment to stimulus values, the additional random process of breaking ties provides the same sampling distribution of equally probable correlation statistics as if there were no ties. The same is true of using this procedure for breaking ties with other randomization tests, also.

Since the introduction of a random procedure for breaking ties provides the same sampling distribution of equally probable statistics as if there were no ties, one need not restrict the application of this tie-breaking procedure to those cases where it is meaningful to think that more subtle measurement would have discriminated between the tied measurements. One can regard this procedure as one that permits the use of published probability tables when the obtained measurements involve ties, whether the ties would still exist with the most sensitive measurement possible or whether more sensitive measurement would eliminate the ties.

The use of random assignment of ranks to break the tie in this example guaranteed a probability of either 3/24 or 1/24. On the other hand, if one followed the practice of discarding tied measurements, he would throw away the two 23s, and would be left with two scores showing a perfect positive rank correlation. The probability of a perfect positive rank correlation with only two pairs of scores, however, is 1/2, which is somewhat larger than 3/24 or 1/24. Nevertheless, in spite of the possibility of getting more significant results with the use of a random process of breaking ties than with discarding tied measurements, some persons are troubled by the deliberate introduction of a random process to gain such an advantage.

There is a way to settle the problem of ties without discarding the tied measurements that will satisfy persons who object to the random process just described. That is to assign the ranks in a way that makes the results least significant. For instance, in the preceding example, the 23 paired with the 16 would be given a rank of 4 and the 23 paired with the 13 would be given a rank of 3, if we were using a one-tailed test for a predicted positive correlation. In this case, without random assignment of the ranks to the tied scores, we have a probability of 3/24.

Another common procedure that is followed is to assign the tied measurements the average rank they would have if each were given a separate rank.

Example 5.16

For instance, if we have measurements 14, 10, 10, 10, and 8, we would give 14 a rank of 1, and each of the 10s would be given a rank of 3, since these three 10s can be considered the second, third, and fourth score in size, although we cannot distinguish between their size. Then the 8 would be given the rank one would expect, 5, since it is fifth in size. In this way, with five ranks, 1, 3, 3, 3, and 5, we have the same sum of ranks as if all five were given ranks 1, 2, 3, 4, and 5. There is, however, no justification for using probability tables derived for untied ranks when we have assigned tied ranks. Consider the effect on rank correlation of substituting the average rank in the case of ties. Suppose we predict a positive correlation and we get the following paired measurements:

Subject	X	Y
a	12	7
b	12	5
c	12	8

We see that there is some variation in Y for the same value of X, but since X has only one value, we could hardly expect to find evidence of a positive or negative correlation between X and Y. Consider, however, what happens when we compute a Spearman rank-correlation coefficient, giving an average rank where there are ties:

Subject	X	Y	$(X-Y)^2$
a	2	2	0
b	2	3	1
c	2	1	1

$$\Sigma\,(X-Y)^2 = 2$$
$$\rho = 1 - \frac{(6)(2)}{3(8)} = .50$$

Instead of a correlation of 0, we have a positive correlation of .50, even though there is no variation in one variable at all. The averaging of ranks in this case led to a bias in favor of a positive correlation.

The curious reader will find, with a little algebraic manipulation, that there is always a Spearman rank correlation of .50 between a set of ranks from 1 to N and another set of numbers each of which has a rank of $(N + 1)/2$. Consequently, all that is needed to get a guaranteed "significant" correlation between any two variables in the world is to hold one variable fixed, take enough different measurements on the other variable, and assign the value of the fixed variable the average of the ranks from 1 to N, for computing the rank-correlation coefficient. A more extreme case yet would be where all observations on both variables were tied and, therefore, the same average rank was assigned to both members of every pair, in which case there would be a perfect positive correlation.

Siegel (1956) suggests that when there are a considerable number of ties one should go through the following steps in computing Spearman's rho:

1. For each X rank for which there are ties, compute $T_x = \dfrac{t^3 - t}{12}$, where t is the number of measurements with that rank.
2. Add the T_x values to get ΣT_x.
3. Perform the corresponding computations for the Y ranks with ties to obtain ΣT_y.
4. Compute $\Sigma x^2 = \dfrac{N^3 - N}{12} - \Sigma T_x$

   and $\Sigma y^2 = \dfrac{N^3 - N}{12} - \Sigma T_y$

   where N is the number of pairs of measurements.
5. Determine the sum of the squared differences between the paired X and Y ranks: ΣD^2.
6. Finally, compute Spearman's rho by the formula

$$\frac{\Sigma x^2 + \Sigma y^2 - \Sigma D^2}{2\sqrt{\Sigma x^2\ \Sigma y^2}}$$

This formula brings the value of rho down to zero in both of the cases just discussed: the case where all values for *one* variable are tied and the case where all values for *both* variables are tied. When there are no tied ranks for either X or Y, the above formula is algebraically equivalent to the usual formula for Spearman's rho, given in a preceding section, "Rank correlation," page 128.

One-subject experiments

Dukes (1965) quite effectively demonstrated through argument and examples of published research the value of studies of only one subject. In spite of the value of such studies, very few experiments published in psychology journals are based on the study of only one subject. This is not surprising inasmuch as editors expect experimental results to be evaluated by statistical tests and there are no handbooks showing how to apply statistical tests to measurements from only one subject. A procedure of statistical analysis of the data from a single subject would be important not only to persons interested in human case studies but also to psychologists like Skinner (1938) and Sidman (1960), who stress the importance of the intensive study of the individual experimental animal.

Perhaps the primary reason statistics books ignore the problem of statistical tests for such data is that the use of only one subject provides no estimate of the variability of the population from which he was selected and consequently no basis for statistical inference about that population. The belief that you cannot statistically generalize to a population of individuals on the basis of measurements from only one subject is certainly well founded. It is also correct, however, that you cannot statistically generalize to a population from which you have not taken a random sample, and this rules out statistical generalization to a population (at least to a population of some importance) for almost all psychological experiments, those with large samples or small.

We have already seen that even without random samples the proper use of random assignment permits us to make valid statistical inferences. The statistical inferences however are restricted to the experimental subjects, generalization to other individuals requiring special empirical assumptions. Statistical inferences also can be made about the effect of experimental treatments on a particular individual when he is *the only subject* in the experiment, provided random assignment is properly performed (Edgington, 1967).

Example 5.17

Consider an experiment with two treatments A and B. We have only one subject. We do not predict which treatment will be more effective, and so we will use a two-tailed test. Twelve treatment times

136 Statistical Inference: The Distribution-free Approach

are written on slips of paper and mixed up in a box. Then six times are drawn from the box and assigned to treatment A, the remaining six times being assigned to treatment B. Our two-tailed null hypothesis is that the treatments have identical effects; that is, the subject will respond the same to each treatment administration as he would if the other treatment had been given at that time. Suppose we obtained the following results listed in the order in which the treatments were given:

$$A \ B \ B \ A \ B \ A \ A \ B \ A \ B \ A \ B$$
$$18 \ 14 \ 15 \ 19 \ 14 \ 19 \ 21 \ 16 \ 20 \ 16 \ 20 \ 17$$

Under the two-tailed null hypothesis, any difference between the mean performance for the two treatments is due solely to the difference in the times at which the treatments were administered. Since the order of administration of treatments is randomly determined, the null hypothesis attributes large differences between treatments to the chance assignment of one treatment rather than the other to various times of administration. To test the null hypothesis, therefore, we consider the sampling distribution of differences between treatments under every equally probable assignment of 6 As and 6 Bs to the 12 obtained measurements, to determine the probability of a difference between treatment means as large as the obtained difference. We can use, for this purpose, the randomization test for a difference between independent samples, with the difference between means as the test statistic. To do this, we first list the measurements separately and compute the obtained difference between means:

A	B
18	14
19	15
19	14
21	16
20	16
20	17
$\Sigma = 117$	$\Sigma = 92$
$\overline{X} = 19\text{-}1/2$	$\overline{X} = 15\text{-}1/3$

$$|\overline{X}_A - \overline{X}_B| = 4\text{-}1/6$$

Our random assignment of treatment times to treatments makes equally probable all 924 (that is, 12!/6! 6!) possible divisions of the 12 treatment *times* between A and B treatments with six times in each treatment. Under our null hypothesis, the response of the subject to a treatment is exactly the same as if he had received the other treatment at that time, and so this hypothesis in conjunction with our random assignment makes equally probable all 924 divisions of the 12 treatment *measurements* between A and B treatments with six *measurements* in each treatment. What proportion of the 924 divisions would provide as large a difference between means as 4-1/6? Notice that there is no overlap of the A- and B-measurement distributions. The division we have is the most extreme one in favor of A treatment. The most extreme division at the other end of the distribution of differences between means is the one in which the six highest measurements are in B and the six lowest in A. The absolute difference between means for that division is of course also 4-1/6. There are, therefore, two out of 924 equally probable divisions that provide such a large difference as the obtained difference. The probability under the null hypothesis is, then, slightly greater than .002. Even at the .005 level, we would reject our null hypothesis in favor of the alternative hypothesis that on at least one of the 12 times at which our subject was given a treatment he would have responded differently if he had been given the other treatment.

Example 5.18

Now let us consider a one-tailed test for the same data, where we predict that the A-treatment mean will be greater than the B-treatment mean. Since we expect the A treatment to sometimes be more effective than the B treatment, we have our null hypothesis assert that the A treatment is *never* more effective than the B treatment for any of the treatment times assigned to our subject. Thus, under our one-tailed null hypothesis, each of the six measurements obtained for the A treatment would have been the same *or greater* if the B treatment had been given at each of those times. And each of the six measurements obtained for the B treatment would have been the same *or less* if the A treatment had been given at each of those treatment times.

The sampling distribution of outcomes under our one-tailed null hypothesis differs from the sampling distribution for the two-tailed null hypothesis because it does *not* imply *identity* of measurement values at a given time for treatments A and B. Consider one of the divisions in the sampling distribution for the one-tailed null hypothesis, that in which the B treatment is assigned the two measurement times which in the experiment had been assigned to A and had given measurements 18 and 21, and that in which the A treatment is assigned the two times that gave measurements 15 and 17 for the B treatment in the experiment:

138   Statistical Inference: The Distribution-free Approach

A	B
15 or less	18 or greater
17 or less	21 or greater
19	14
19	14
20	16
20	16
$\Sigma = 110$ or less	$\Sigma = 99$ or greater
$\overline{X} = 18\text{-}1/3$ or less	$\overline{X} = 16\text{-}1/2$ or greater

$$\overline{X}_A - \overline{X}_B = 1\text{-}5/6 \text{ or less}$$

Under our one-tailed null hypothesis, the only *exact* response measurements we can specify for this division are for those times which are assigned to the same treatment as in our actual experiment. These are the measurements which do not have "or less" or "or greater" after them. The other response measurements do not have any *exact* values under the one-tailed null hypothesis, and so all we can do is give the measurement actually obtained at that treatment time and add "or less" or "or greater."

Notice that this division would under the *two-tailed* null hypothesis be exactly the same, except for the absence of the qualifiers "or less" and "or greater." Thus, each of the 924 equally probable divisions, under the *two-tailed* null hypothesis, can be converted to a division under our *one-tailed* null hypothesis by adding "or less" to some of the measurements listed under A, or "or greater" to some of the divisions listed under B. It follows that under our one-tailed null hypothesis we have exactly the same distribution of $\overline{X}_A - \overline{X}_B$ as under the two-tailed null hypothesis, except for the qualifying "or less" added to the value.

We can therefore say that the probability under the one-tailed null hypothesis of getting such a large difference between means in favor of A as 4-1/6 is no greater than 1/924, or approximately .001, because the probability cannot be greater under our stated one-tailed hypothesis than under the two-tailed hypothesis.

Inspection of the measurements in the sequence A B B A B A A B A B A B, which was presented at the beginning of Example 5.17, reveals a tendency within each type of treatment for the later measurements to be larger, the same sort of tendency that is associated with practice effects in the early stages of learning. The presence of this tendency has no effect on the validity of the significance test because the random assignment of treatments to times of administration randomizes any beneficial or detrimental effects on performance associated with the time of administration, and the significance test takes this randomization into account.

When the treatments are given so close together that the response to a treatment is affected by the preceding treatment administration, we cannot make statistical inferences about temporally isolated treatments. This problem also exists with large samples and, as Lindquist (1953) and Winer (1962) indicate, is not solved by using all possible orders to "balance out" the effect of earlier treatment administration. For example, if a B treatment enhanced the response to the subsequent A treatment more than an A treatment did the response to the subsequent B treatment, averaging over all orders would tend to yield a significant difference in favor of the A treatment in instances showing no treatment differences with temporal isolation of treatment administrations.

Another source of difficulty in interpreting a significant difference between treatments for one individual, in some experiments, is the possibility that the obtained difference is the result of the individual's attitude or expectation regarding the experiment. The same problem exists for large samples. In a study using several subjects, for example, there is no reason to assume that the directional effect of the attitudes varies from subject to subject in such a way that the overall average effect is zero; in fact, in many cases it is reasonable to assume a similar direction of bias in almost all of the subjects. We should try to eliminate this source of error—for example, by concealing the purpose of the experiment or by using naive subjects.

In an experiment employing only one subject, the statistical inference that the treatments had a differential effect refers to the experimental subject alone—for at least one of the treatment times. To generalize to other individuals requires nonstatistical reasoning. We have argued that in the experiments involving hypothesis testing we want to generalize to individuals of a particular kind, not to an existing population of heterogeneous individuals. For this kind of generalization we need to consider which subject variables or characteristics are likely to be relevant to responding differentially to the treatments. Certain characteristics may be known to be relevant and others known to be irrelevant, but there are usually a number of characteristics whose relevance is unknown. For example, in a sensory perception experiment, the sex, intelligence, and visual acuity of the subject are potential determiners of whether the subject will respond the same to one treatment as the other or whether he will respond differentially to the treatments. Generalizing from an experiment employing only one subject can be especially difficult, since our subject has only one sex, level of intelligence, and degree of visual acuity.

Aspects of experimental design to facilitate nonstatistical generalization

So far the discussion of randomization tests has emphasized procedures for setting up experiments in a way that permits making statistical inferences about differential treatment effects on the subject or subjects actually used in an experiment. It is equally important to consider, in the design of an experiment, how to make *nonstatistical* generalizations to other individuals.

Consider the problem just stated in regard to one-subject experiments—the fact that we cannot statistically generalize to any other subjects, and that the uniqueness of a single subject with regard to his characteristics may make it difficult to generalize even on a *nonstatistical* basis. How can we determine which characteristics possessed by our subject are relevant in generalizing and which are not?

In order to get evidence regarding the relevance of certain characteristics' responding differentially to the treatments, we can carry out replications with other subjects, obtaining enough measurements from each subject to determine the significance of the difference between treatments for each subject. For these replications we should select subjects that differ with regard to one of the characteristics under investigation. For example, if our first subject in the perception experiment was a male with poor vision and high intelligence, we should replicate with a female with poor vision and high intelligence, a male with good vision and high intelligence, and a male with poor vision and low intelligence. Of course, if we want to consider the possibility of these characteristics interacting in their influence on the experimental outcome, we should also replicate with other combinations of sex, vision, and intelligence.

It is not always possible to carry out a series of replications of an experiment by using only one subject per replication. There are situations in which it is necessary to have more than one subject in order to get the measurements needed for a significance test. For example, treatments may be mutually exclusive, such as raising an animal in darkness or raising it under normal conditions of lighting, making it impossible to give the same subject both treatments. For such experiments, we randomly assign a number of subjects to each experimental treatment. The statistical test is carried out in the same way as with repeated measurements from one subject, and the null hypothesis tested is that every measurement is the same as it would have been if the other treatment had been given instead of the treatment actually given. Rejection of the null hypothesis implies acceptance of the hypothesis that the measurements are not the same as they would have been under the other treatment, but since we cannot identify the measurements which would have been different, we cannot determine which subjects would have responded differently. Consequently, if our subjects differ considerably from each other with regard to several characteristics, we cannot have a very definite idea of the type of individual to which we can generalize. In order to avoid this ambiguity, we should select subjects who are homogeneous with regard to the characteristics whose relevance we want to investigate. Suppose our first group of subjects were young males and that we obtained a significant difference between treatments. We could then determine whether age and sex are important variables by replicating the experiment with a group of old males, a group of young females, and a group of old females, each of the groups being statistically analyzed separately.

We have just stressed the value of *maximum homogeneity* of the experimental subjects in making nonstatistical generalizations. There is an alternative view, however, in favor of *maximum heterogeneity*. Since the view seems to be inconsistent with the general attitude of this book regarding the value of intensive, detailed analysis, it is *not* recommended. Nonetheless, it is an interesting alternative, worth a brief discussion.

The argument in favor of maximum heterogeneity hinges on the fact that a significant difference between treatment effects is not likely to occur when only one or a very few individuals are differentially affected, even though the *statistical* inference refers to "at least one" individual. If it can be assumed that several subjects with differential effects are likely to be required to provide a significant difference between treatments, then a significant result implies that in a heterogeneous group there would have to be several types of persons responding differentially to the treatments. Significant results with a homogeneous group would not lead to such an implication.

In order to contrast the view favoring maximum homogeneity with the view favoring maximum heterogeneity, let us consider a hypothetical example. For the sake of illustration, suppose that age is the only subject attribute that conceivably could influence whether a subject responded differentially to treatments A and B. The experimenter favoring homogeneous groups decides to use subjects who are 80 years of age and the experimenter favoring heterogeneous groups uses subjects ranging from 10 to 90 years. Each experimenter gets significant differences between treatments *A* and *B*.

The experimenter with the homogeneous group can state that 80-year-old persons would respond differentially to treatments *A* and *B* and be rather sure of being correct in this statement, which refers to a very narrow class of individuals. The experimenter with the heterogeneous group has some basis for making the inference that both young and old persons respond differently to treatments *A* and *B*, but he cannot be as certain of this inference as the other experimenter can be of his inference about a much narrower class of individuals.

The choice between making groups maximally homogeneous or maximally heterogeneous depends, then, on whether one prefers a sure statement about a very narrow class of individuals or a less certain statement about a broad class of individuals.

The design of longitudinal experiments

Measurements occur later than treatments, and so all experiments are extended over time, but there are experiments which are extended over such long periods of time that it is especially important to allow for variation in the subject over that time period.

For example, in longitudinal studies we may study the same children over a period of several years. Most longitudinal studies are nonexperimental, but experimental longitudinal studies can be carried out. We shall examine the design of a longitudinal experiment on an individual. Reasons for performing an experiment on an individual rather than a group of individuals were given in the two preceding sections. Replications of longitudinal experiments should be planned according to the recommendations in those sections.

It should be stressed that we will be considering the design of *experiments* for the longitudinal study of individuals, not the usual research with case studies. In sociology and medicine, for example, case studies are rather commonly used, but the assessment of the influence of certain variables in the life of the individual is usually a qualitative one. Even when quantitative data are gathered, seldom do they permit statistical inferences about the influence of a particular variable.

We will consider the administration of an experimental treatment at one or more randomly selected times, the results of which are to be compared to measurements when no treatment is given. It is desirable to have several experimental treatment administrations, because increasing the number of experimental measurements increases the sensitivity of the test rapidly. For instance, with just one experimental treatment and five control treatments, the probability, under the null hypothesis, of getting the largest measurement from the experimental treatment is large: 1/6. But with only two more experimental treatments a small probability is possible; the probability that the experimental measurements are the three largest is 1/56.

Example 5.19

Now let us consider in detail how to set up an experiment whenever it is necessary to take several measurements on a subject over a fairly long period of time. First we select our subject. Then we determine certain possible times at which the experimental treatments could be administered. Suppose 10 consecutive Mondays are the times from which we will select our treatment times. We draw five of these dates from a hat and use them for our experimental treatment days. The other five Mondays we do not administer the treatment, but allow the subject to carry on his normal daily activities. On each of the five treatment days we provide an experimental treatment—for example, electroshock, psychotherapy, drugs, or physical exercises. On Tuesday, the day after the treatment, we take a measurement of an attribute that we expect to be affected by the treatment, such as the subject's blood pressure, his anxiety, or his intelligence. Also, on the Tuesday after the days when no treatment was administered, measurements of the attribute are made. Then we use a statistical test to determine whether the difference between these two sets of measurements is statistically significant. The null hypothesis is that the response of the subject on each of the 10 Tuesdays was not affected at all by the treatment; that is, (1) the responses on each of the Tuesdays following a day when a treatment was given were exactly the same as they would have been if the treatment had not been given, and (2) the responses on each of the Tuesdays following a day when the treatment was *not* given were the same as they would have been if the treatment had been given. Rejection of the null hypothesis implies that it did make a difference on at least one of the Tuesdays whether the treatment was or was not administered on the preceding day.

Now, of course, one cannot draw statistical inferences about the effect of a treatment itself if each time it is administered there are a number of other things associated with it, such as the presence of a nurse, or the dispensary background with its antiseptic smell. The statistical inference simply concerns the effect of everything associated with a Monday being a treatment day or not being a treatment day. To make it more specific requires empirically based arguments about the irrelevance of systematic differences between treatment days and nontreatment days.

If there are lingering effects of the treatment, obviously we want the times when we measure the effect of the treatment and those when we measure the effect of the "nontreatment" to be far enough apart to prevent artificial inflation of the nontreatment measurements. On the other hand, we do not want the measurements taken too far apart, or there will be so much variation within our treatment measurements and within our nontreatment measurements over that time that it will be difficult to detect a difference. The same sort of difficulties may arise in taking a large number of measurements over a long period of time rather than a small number over a short time. The sensitivity of the statistical test may be reduced more by the heterogeneity due to variation over a long period of time than it is increased by the added number of measurements.

Example 5.20

When there is occasion to use repeated measurements of effects of experimental and control conditions over a period of time during which it can be expected that there will be a rather consistent increase or decrease, or fairly long cycles, a simple comparision of measurements under control and experimental conditions is not likely to be sensitive. It would be better to use a *change measurement,* that is, a measure of change. For instance, suppose we believed that even without the experimental treatment our subject would improve noticeably over the time period of our experiment. Then we would expect our distribution of Tuesday measurements following experimental days to overlap the distribution of Tuesday measurements following control days because of the consistent increase in both types of measurements over the time period of the experiment. This would make our test for differences rather insensitive. What we should do is use change measurements. We only administer our treatments on the five Mondays randomly assigned to treatments, but on every Friday we obtain the same kind of measurement as on Tuesday. Then we subtract each Friday measurement from the following Tuesday measurement to obtain a change measurement. This change measurement is assigned to the condition (that is, treatment or nontreatment) associated with the Monday enclosed within the Friday-to-Tuesday time span. Then we carry out our statistical tests with the change measurements. Under the null hypothesis, with our random assignment, it is equally probable that each change measurement is assigned to the treatment or nontreatment group. Thus a randomization test for a difference between independent samples can be carried out on the change measurements to determine whether to reject the null hypothesis of no treatment effect. Simple change measurements can be useful when there is reason to expect a linear increase or decrease in measurements, but when we expect the amount of change to be proportional to the measurements, it would be better to use the difference between the *logarithms* of the measurements as the change measurement because a proportional increase in measurements provides an arithmetic increase in the logarithms of the measurements. If logarithms were not used and there was a proportional increase, there would be considerable variability of change measurements within both the treatment and the nontreatment groups, providing considerable overlap of the distributions, and thereby making it difficult to detect a difference.

Suppose there is reason to expect that the treatment, if it has any effect at all, will have a permanent effect and thereby will cause a consistent increase in the measurements for both control and experimental sessions. Then after the first few treatments the subsequent treatments may have no effect, which would weaken the statistical test. When the treatment is assumed to have a permanent effect, it is desirable for experimental purposes to give a dosage expected to raise the level slightly each time, so that a number of treatments can be administered before reaching the stage where additional treatments have no effect. At the time of the experiment we do not know whether a treatment will have any effect at all, but even guesses about the optimal dosage may be helpful in detecting its effect if it has any.

Example 5.21

If we have a fairly definite idea of the amount of time required for a treatment to show its full effect, if it has any effect at all, we can carry out a fairly sensitive experiment where only one treatment administration is possible, and this administration produces a permanent change in measurement. Suppose we had reason to believe that the treatment will have its maximal effect in one week, if it has any effect at all. Then we can randomly select one Monday from the 52 Mondays in the following year as the treatment day. All of the other Mondays are control days. We take measurements on every Monday, both those before and those after the treatment. Then we work up change measurements by subtracting the measurement for each Monday from that for the following Monday. The treatment-condition change measurement is the one involving the Monday the treatment was given and the following Monday. The other 51 are control-condition change measurements. Suppose the treatment-condition change measurement was the largest of the 52. Under the null hypothesis, the probability of this is 1/52.

Statistical independence

We cannot carry out a statistical test with ten measurements from each of five individuals as we would with one measurement from each of 50 individuals. The usual explanation is that repeated measurements from an individual are "correlated measurements" and are, in some sense, not statistically independent measurements. But the section "One-subject experiments" showed that a special use of random assignment permits 50 measurements from one individual to be analyzed in the same way as one measurement from each of 50 individuals. In this case random assignment made the measurements statistically independent. To further examine the role of random assignment in statistical independence of measurements we will now consider some examples where repeated measurements are *not* statistically independent.

Example 5.22

Consider Figure 5.1, which is a plot of the mean errors over ten learning trials for two groups of rats randomly assigned to two treatments. One cannot tell from the graph, of course, whether the two groups

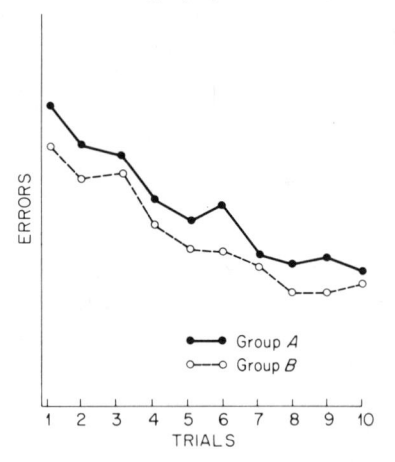

Figure 5.1 Mean errors of two groups of rats over ten trials.

differ significantly on any trial, but the fact that group A made more errors than group B on all ten trials suggests that this is not a chance difference. But what is meant by saying that it is not a chance difference? If anything more than a vague feeling is involved, more than likely something like this is implied: If one were to toss a coin on each of the ten trials to determine the group to be "assigned" the most errors, the probability that one of the two groups would be assigned the most errors on all trials would be $2/2^{10}$, or approximately .002. Suppose we had assigned at random one of a pair of rats to treatment A and the other to treatment B and measured the number of errors on *one* trial. Then we randomly assigned a different pair of rats and determined their errors on *two* trials, and so on for 10 pairs of rats. Then, indeed, if we got the graph shown in Figure 5.1, we would conclude that the treatments did not have the same effect. However, the different points on this graph do not represent different animals but the same animals. The consistent superiority of group B over all trials could be solely the result of the assignment of better learners to that group, since even with random assignment of subjects to groups we rarely get exactly equivalent groups. If we had used ten different groups of experimental and control animals for the ten trials and obtained the same kind of graph, we would have had evidence that learning was better under one experimental treatment than under the other because the ten trials would have been made by different animals. But, since we used the same animals over all trials, we have correlated performance: the animals that performed well on one trial performed well on other trials.

Consistently better performance over a number of tests could result from individual differences even if groups A and B were assigned to *identical* experimental treatments. An example involving human subjects will illustrate this point. We randomly divide a class of 100 students into two groups of 50 students each. Then we administer an intelligence test to the 100 students. Group 1, say, got a slightly higher mean score than group 2. Now we give the same groups a series of very similar intelligence tests, tests that include similar items or which test similar abilities. With high reliability of the test scores, we could easily get higher means for group 1 on every one of a dozen similar tests. Such consistent superiority of group 1 certainly could not indicate that the difference in intelligence scores resulted from differences in experimental treatments because the treatments were identical. The consistent superiority of group 1 would reflect differences in the average intelligence of the two groups, a difference which is not uncommon in random assignment of subjects to two groups. Random assignment of subjects is not a guarantee of identity of groups but a guarantee of being able to statistically take into account individual differences.

Example 5.23

Instead of using a significance test or a graph to show the effect of an experimental treatment, sometimes photographs or sketches are published. For example, suppose that Figure 5.2 represents the cartilage cells from an animal's nose, before and after the animal was given a certain disease. The pictures suggest that the disease made the cells more heterogeneous in size. The pictures, however, can be misleading because here, as in the previous example, the reader is likely to be making comparisons on the basis of correlated observations which are treated as if they were independent.

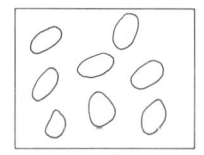

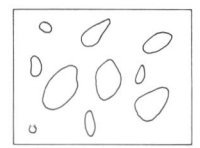

Figure 5.2 Cartilage tissue slices before (left) and after (right) disease.

There are 18 different cells in the two slices but we know that they were not randomly assigned to the disease and nondisease conditions because diseases cannot be "assigned" to certain cells in a nose and kept away from others. Thus, there are *two* sets of correlated measurements, not *18* independent measurements. On the other hand, one measurement from each of 18 randomly assigned *animals* would provide 18 statistically independent measurements of cell size variability that could be used in a statistical test of treatment effects.

But suppose we were simply interested in whether the homogeneity of sizes of cells after the disease differed from the homogeneity before the disease, without attributing such a difference to the disease. We would still have difficulty, because we have pictures of neither eight cells selected at random from the nose before the disease, nor 10 taken at random after the disease. At best we have *two* randomly selected *slices* of tissue from the same nose, one taken before the disease, the other after. For all we know, the difference between these two slices of tissue may be no greater than between random slices before the disease or between random slices after the disease. The only way to get an idea of the variability among slices is to take a number of slices at random from the animal before and after the disease. A single slice of tissue is not comparable to a sample of liquid in the sense that one slice is just like another.

Example 5.24

Figure 5.3 is conducive to the same kind of misimpression as Figure 5.2. Here we have a difference in the clumping of cells under two treatments. In this case it may be that the two animals *were* randomly assigned to treatments K and R, but regardless of the number of cells on the two slides, we have only one measurement from each animal, and so we cannot carry out a statistical test for the difference in the effects of the treatments. To put it differently, under the null hypothesis of no difference in treatment effects, the probability is 1/2 that the animal given treatment K would show greater clumping.

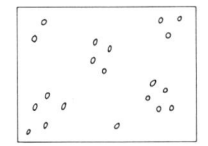

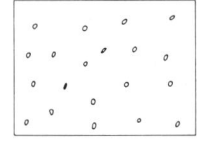

Figure 5.3   Tissue slices for treatment K (left) and treatment R (right).

Consider the main features of a randomization test for a difference in cell clumping. The null hypothesis is that under the alternative treatment each individual would have provided the same slice of tissue with the same pattern of cells that he did in the experiment. A measure of clumping for a tissue slice could be the distance, averaged over the cells, of each cell from its *nearest* adjacent cell in the slice. The test statistic could be the sum of the clumping measurements for the A treatment slices.

Example 5.25

The concept of statistical independence is automatically taken into consideration in performing a randomization test. The appropriate divisions or pairings of data are determined by the null hypothesis and the random assignment. *Measurements* are never randomly assigned; subjects are randomly assigned and, under the null hypothesis of identical treatment effects, take their individual measurements or sets of measurements with them to the treatment they are assigned to. If there are five subjects randomly assigned to treatment A and five to treatment B, there are 10!/5! 5!, or 252, equally probable divisions of the 10 subjects for the randomization test. The null hypothesis of no difference between treatments implies that each of the 10 subjects takes his measurement (or set of measurements) to the treatment he is assigned to, and so there are 252 equally probable divisions of the obtained measurements of the 10 subjects. But notice that it is the *subject* that is randomly assigned to one treatment or another, and that if each subject was measured 20 times, there are *not* 200 measurements divided in various ways, but rather there are 10 *sets* of 20 measurements to be divided in every way into five sets for treatment A and five for treatment B. Ordinarily, one would reduce each set to a single mean, and divide the 10 means as though they were individual measurements, which is consistent with the null hypothesis and the random assignment procedure. No matter how many measurements are made on each subject, if there have been only five subjects randomly assigned to one treatment and five to the other, there are only 10 statistically independent numbers (or sets of numbers) that can be obtained.

On the other hand, with only one subject, and random assignment of treatment times, say 10 to treatment A and five to treatment B, there are 15!/10! 5! or 3,003 equally probable divisions of the treatment times among treatments A and B. Under the null hypothesis, then, there are 3,003 equally probable divisions of the measurements obtained for the 15 treatment times. Now the reason we have 15 independent numbers with one subject, and only 10 for the 10 subjects, is that the independence is determined by the random assignment procedure.

In Chapter 3 it was noted that cluster sampling does not provide as precise an estimate of a population mean as does a simple random sample with the same number of individuals. The reason is that every individual in a simple random sample is independently selected and therefore provides a statistically independent measurement whereas for cluster sampling only the clusters are independently selected, not the individuals within them.

Experimental independence

In the preceding section we saw the role that random assignment plays in determining statistical independence. In this section we will see that even when there has been random assignment adequate to justify the assumption of a given number of independent measurements, there may be uncontrolled experimental variables which make the measurements *interdependent*.

Example 5.26

Lack of independence of treatment effects among individuals can complicate or even invalidate statistical analysis. To illustrate, take a comparison of the proportion of vaccinated and unvaccinated animals that catch the disease which the vaccine is intended to prevent. One hundred animals are randomly assigned to be vaccinated and one hundred others not to be vaccinated. During the course of the experiment, 20 of the 100 unvaccinated animals and none of the 100 vaccinated animals catch the disease. The difference is impressive, but it is easy to misinterpret. Suppose that the vaccinated and unvaccinated animals were given identical care *except* for being kept in separate buildings and being attended by different laboratory assistants. If one animal in the unvaccinated group got the disease, it could accidentally be spread to the other unvaccinated animals by the laboratory assistant through handling of the animals. It is possible that if one of the *vaccinated* animals had caught the disease, it would have spread to the other vaccinated animals. Thus the difference in the disease rates for the two groups could be solely the result of a single animal being assigned to the control treatment rather than the vaccination treatment.

There are numerous studies in which it is difficult to get independent treatment effects for the individuals. For example, two equally competent teachers could have classes with considerably different amounts of knowledge about a subject simply because one of the teachers had a student who asked provocative questions.

Similarly, when experiments are carried out on groups of persons to find out the effect of isolation of the group from society, of semistarvation, or of other factors, we do not know to what extent one or two persons essentially determine the responses of the entire group. It does little good to have a large number of independently assigned *individuals* when the *measurements* are not independent.

The problem of *interdependent* measurements from independently selected subjects was discussed in regard to estimation in the section "Reliability of measurements" in Chapter 4. The recommendation made there is also applicable to hypothesis testing: Precautions should be taken to ensure that each individual provides the same measurement as if he were the only individual in his group.

Approximate randomization tests

The principal disadvantage of randomization tests is the great amount of computation required. For example, to determine the significance of a correlation between two sets of 10 measurements, it is necessary to compute 3,628,800 correlation statistics because there are 10! different ways of pairing two sets of 10 measurements. Even an electronic computer that paired the measurements and computed a correlation statistic every second would take over a month of continuous operation to compute 3,628,800 statistics.

The amount of computation for a randomization test can be lowered to a practical level by using random samples of all of the pairings or divisions in the entire sampling distribution to obtain a smaller sampling distribution (Edgington, 1969). We will use the term *approximate sampling distribution* to refer to this smaller distribution of pairings or divisions (and to the distribution of test statistics associated with them) obtained by randomly sampling the entire sampling distribution. The term *approximate randomization test* will refer to a randomization test that uses an approximate sampling distribution instead of the entire sampling distribution.

How do we determine the significance of an obtained statistic by means of the approximate randomization test? Take as an example the correlation problem just referred to. First we compute a correlation statistic for our obtained data. Then we have an electronic computer use a table of random numbers to simulate taking a random sample, *with replacement,* of, say, 999 pairings from the entire sampling distribution of 3,628,800 and compute a correlation statistic for each of the pairings. This provides us with an approximate sampling distribution of 999 correlation statistics in a matter of minutes.

Next we determine how many of the 1,000 correlation statistics (the 999 in the approximate sampling distribution plus the obtained statistic) are as large as the obtained statistic. This number is then divided by 1,000 to get the probability, under the null hypothesis, of such a large correlation statistic.

The justification for this procedure of determining the significance will now be given. Random assignment in conjunction with the null hypothesis permits us to regard our obtained *pairing* of data as a random sample of one pairing taken from a distribution of 3,628,800 pairings. Consequently our obtained *correlation statistic* can be regarded as a random sample of one correlation statistic taken from the distribution of 3,628,800 (not necessarily all different) correlation statistics associated with the entire distribution of pairings. When we add our obtained correlation statistic to the 999 correlation statistics that were randomly selected, we have, under the null hypothesis, 1,000 randomly selected correlation statistics from the entire sampling distribution. We now consider the probability that the first-selected statistic (the obtained statistic) will be the largest of the 1,000. This is the type of problem previously dealt with in the section in Chapter 4 entitled "Distribution-free confidence interval for an individual."

If all 1,000 statistics had different values, so that they could be ranked from high to low with no ties, under the null hypothesis the probability would be 1/1,000, or .001, that our obtained statistic will have any specified rank from 1 to 1,000. Thus the probability that it will be the largest of the 1,000, having a rank of 1, would be .001. Admitting the possibility of ties, the probability is *no greater than* .001 that our obtained correlation statistic will be larger than all of the statistics in the approximate sampling distribution. Similarly, the probability is no greater than .002 that our obtained statistic will be one of the two largest of the 1,000, and so on. This, then, is the justification for asserting that the probability, under the null hypothesis, is the proportion of the 1,000 values that are as large as the obtained value. (We could have used the same argument if sampling *without* replacement had been carried out, but our subsequent discussion of the power of the approximate randomization test will be considerably simplified by restricting it to distributions obtained by sampling *with* replacement.)

Approximate randomization tests are therefore *valid, not approximately valid,* procedures for testing null hypotheses. For example, when the null hypothesis is true, there is a probability no greater than .05 of rejecting it if it is rejected only when the obtained statistic is in the upper 5 percent of the distribution consisting of the approximate sampling distribution plus the obtained statistic. Probabilities computed from approximate sampling distributions are not simply approximations to probabilities based on the entire sampling distribution but are probabilities in their own right.

Next we will consider the relative power of the approximate randomization test based on an approximate sampling distribution of 999 statistics. We will first prove the following statement:

> Statement 1: The probability is .99 that no more than 17 statistics from the upper 1 percent of the entire sampling distribution will be in the approximate sampling distribution.

It will then be shown that statement 1 implies statement 2:

Statement 2: The probability is .99 that an obtained statistic that would be judged significant at the .01 level by using the entire sampling distribution will be given a probability no greater than .018 by using the approximate sampling distribution.

To prove statement 1, we regard the entire sampling distribution as consisting of two types of statistics: those that are in the upper 1 percent of the distribution and those that are not. In randomly sampling the entire sampling distribution with replacement, for a single draw there is a probability $p = .01$ of drawing a statistic from the upper 1 percent and a probability $q = .99$ of not doing so; that is, of drawing a statistic from the lower 99 percent of the distribution. Replacement after each draw ensures that p and q remain constant over all draws. The binomial distribution of the number of statistics from the upper 1 percent of the entire sampling distribution for $N = 999$, $p = .01$, and $q = .99$ is almost normally distributed, with $\mu = Np = 9.99$ and $\sigma = \sqrt{Npq} = 3.14$. Ninety-nine percent of the area under a normal curve falls below a point that is 2.33 standard deviations above the mean. Thus, the probability is .99 that within the approximate sampling distribution, the number of statistics from the upper 1 percent of the entire sampling distribution is no greater than $9.99 + (2.33)(3.14) = 17.31$. This verifies statement 1 since in the continuous normal distribution the discrete frequency 17 is assumed to occupy the interval 16.5 to 17.5.

Computations that have been made by an electronic computer on the *exact* binomial distribution (*not* the normal curve approximation) have also verified statement 1.

In order to see that statement 2 follows from statement 1, it is necessary to appreciate that statement 1 is true regardless of the magnitude of the obtained statistic. This is so because the selection of the statistics in the approximate sampling distribution is in no way affected by the magnitude of the obtained statistic. Statement 1 therefore is true even when the obtained correlation statistic is large enough to be judged significant at the .01 level by means of the entire sampling distribution. This fact enables us to derive statement 2 from statement 1, which we will now do. To say that our obtained correlation statistic would be judged significant at the .01 level by means of the entire sampling distribution is to say that it has a value that lies within the upper 1 percent of the entire sampling distribution. The only statistics in the approximate sampling distribution that could possibly be as large as the obtained statistic therefore would be those from the upper 1 percent of the entire sampling distribution. This fact, in conjunction with statement 1, leads us to this conclusion: When the obtained statistic would be judged significant at the .01 level by means of the entire sampling distribution, the probability is .99 that no more than 17 of the values in the approximate sampling distribution will be as large as the obtained statistic. When no more than 17 statistics in the approximate sampling distribution of 999 are from the upper 1 percent of the entire sampling distribution, no more than *18* (17 in the approximate sampling distribution plus the obtained statistic itself) of the 1,000 will be as large as the obtained statistic. Statement 2 therefore must be true.

Statement 2 is correct whether the null hypothesis is true or whether it is false, whether there is a treatment effect or whether there is not. We are, however, particularly interested in the application of statement 2 when the null hypothesis is *false* because it shows that the approximate randomization test using an approximate sampling distribution of 999 statistics is almost as powerful as a randomization test using the entire sampling distribution.

Similar computations and reasoning in regard to the binomial distribution of the number of statistics from the upper 5 percent of the entire sampling distribution for $N = 999$, $p = .05$, and $q = .95$ lead to this statement:

> Statement 3: The probability is .99 that an obtained statistic that would be judged significant at the .05 level by using the entire sampling distribution will be given a probability no greater than .066 by using the approximate sampling distribution.

Statement 3, like statement 1, has also been verified by electronic-computer computations on the exact binomial distribution.

Statements 2 and 3 apply only when the approximate sampling distribution consists of 999 statistics selected at random, with replacement. The validity of the statements is, however, independent of the size of the entire sampling distribution since the sampling is with replacement. Whether there are 50 thousand or 50 trillion statistics in the entire sampling distribution is irrelevant. Statements 2 and 3 apply, of course, to any randomization test statistic, not just to correlation statistics.

Transformations

Stevens's (1951) views regarding the relationship between scales of measurement and statistics have been controversial, but there can be no argument about the strong influence he has had in making psychologists think about measurement. He has taken a commendable stand against the blind manipulation of numbers performed on the off-chance of discovering something useful or meaningful.

The approach in this book has followed the spirit, if not the letter, of Stevens's recommendations regarding measurement and statistics. For example, examination of the purpose in estimating a mean or a total led directly to considerations of the meaning of the quantity being estimated. His concept of the invariance of relationships under various mathematical transformations has been especially useful. The lack of invariance of the mean under nonlinear transformations (except for the unique case of the mean of one measurement) was used to show that the nonlinear transformations that normalize also make it impossible to estimate the mean of a population in the original units of measurement. In this section we will consider further implications of the lack of invariance of the mean under nonlinear transformations.

Example 5.27

Speed of performance can be expressed in terms of amount performed per unit of time or amount of time per unit of performance. The following example (Edgington, 1960a) will show how different the interpretation of a difference between means can be, for these two alternative measures of speed of performance.

The speeds shown in Table 5.1 are hypothetical running speeds for two rats, each rat being measured on two different occasions. The same measurements are expressed in two ways in the table. The table shows that rat *A* on the average traveled more feet per second and was, therefore, *faster*. However, the *same rat*, rat *A*, also on the average took more seconds per foot traveled and was therefore *slower*.

Table 5.1 Speeds of rats expressed in two ways

	Feet per second		Seconds per foot (1/feet-per-second)	
	Rat *A*	Rat *B*	Rat *A*	Rat *B*
	2	4	.50	.25
	8	5	.125	.20
Total:	10	9	.625	.45
Mean:	5	4.5	.3125	.225

One way of settling the contradiction would be to accept the convention of physicists to express velocity only in terms of distance per unit of time but this solution would not satisfy most researchers, who consider distance per unit of time and time per unit of distance to be equally valid expressions of speed. And in fact physicists themselves often determine velocity by measuring the time taken to cover a given distance.

Frequently, in research the latency times and running times of rats are subjected to a reciprocal transformation before statistical analysis; that is, the statistical analysis is performed using the reciprocals of the running times and the reciprocals of the latency times. However, a difference between the means of the reciprocals of running times may be in the opposite direction from the difference between the means of the running times, as we have seen.

Up to this point the discussion has dealt with running times and latency times of rats, and the reciprocals of these measures. The statistical contradiction presented, however, is not the result of carrying out the statistical analysis with the reciprocals instead of the original data. The two reciprocally related measures are equally valid measures of speed of performance, and either one could be considered to be the reciprocal of the other. There are simply two precise measures of speed of performance that can lead to contradictory conclusions, for certain statistical computations. This fact applies to speed of performance of any kind. It applies, for example, to measuring manual dexterity with a pegboard in terms of amount of time per peg or in terms of number of pegs per unit of time. Another example is the number of nonsense syllables memorized per unit of time compared to the amount of time per nonsense syllable memorized. And verbal as well as nonverbal tests can employ as a measure of speed either amount performed per unit of time or amount of time per unit of performance. In these and other cases where speed of performance is measured, the amount of obtained difference and its direction can be inconsistent with what would have been obtained if the reciprocally related measure had been used.

The preceding example appears paradoxical only because of a tendency to think of means as measurements. When rats travel at a constant speed, a rat that travels more feet per second than another rat takes less time per foot traveled; therefore, there is a tendency to also believe that a rat that travels, *on the average,* more feet per second must take less time per foot traveled, *on the average.* In other words, the means are implicitly regarded as having the same sort of relationship as the individual measurements. One does not, however, observe the same sort of relationship in a comparison of two *groups* of individuals as in a comparison of two individuals. We can see that one rat is running faster than another, because the distance between them is increasing, but we cannot directly observe distances between *means* changing while two groups of rats compete. The mean is a *statistic, not a measurement,* and a relationship between two means does not remain invariant under all the transformations of measurements that leave the same relationship between two measurements invariant.

Example 5.28

An example of a different type (Edgington, 1960b) will be given to further stress the importance of recognizing that a mean is not a measurement. In experimental situations there may be two dimensions which are so related that one dimension absolutely determines the other. If we assume circularity of the pupil of the eye, for example, then the pupil diameter and pupil area are such dimensions. They are related by the formula $A = 1/4\pi d^2$, and so anything that affects one dimension must affect the other. Consider a hypothetical case in which the experimenter measures both the pupil area and the pupil diameter to determine the effectiveness of pupillary conditioning. The following table shows the pupil sizes for two subjects, expressed in terms of pupil diameter and pupil area, under two experimental conditions:

Table 5.2 Pupil sizes under two conditions

	Condition A		Condition B	
	Pupil diameter	Pupil area	Pupil diameter	Pupil area
	2.8	6.2	5.1	20.4
	7.0	38.5	5.1	20.4
Total:	9.8	44.7	10.2	40.8
Mean:	4.9	22.3	5.1	20.4

The mean pupil size expressed in terms of pupil *diameter* is *smaller* on the average for condition A than for condition B but is *larger* on the average in terms of pupil *area.*

Conditioning of the pupil should be manifested equally well in the pupil diameter and the pupil area, and so these two dimensions should be equivalent for the purpose of testing a hypothesis about pupillary conditioning, but any conclusions based on a difference between means can be affected by the choice of the dimension to measure.

In the chapter on estimation, in the discussion of the practical purposes of estimating means and differences between means, it was suggested that estimates of these parameters are of value because they can be converted to estimates of totals. Now two alternative measures or dimensions may be equally appropriate for a comparison of means but not for an estimation of a total. The following brief consideration of this point throws light on the paradoxes in Examples 5.27 and 5.28.

Let us consider an example of a situation similar to that involving the paradox associated with diameters and areas, where we measure the diameters and *volumes* of ball bearings. Should we estimate the mean diameter or the mean volume? It depends on the total we are interested in or what totals we want to compare. If the total diameter of all the ball bearings, laid side by side, is of interest, we would estimate the mean diameter. If on the other hand we were interested in the total volume (to convert it to an estimate of the total weight?), the mean volume would be the statistic of interest.

Thus we see that specification of purpose can make one of two or more "contradictory measures" more appropriate than another, thereby resolving the paradox . . . at least for *estimation.* If indeed we were trying to estimate the difference between two populations of circles, there would be no inconsistency in concluding that *population A's mean diameter* was smaller than population *B*'s and that *population B's mean area* was smaller than population *A*'s because both conclusions could be simultaneously correct.

Now suppose that our objective is not the provision of estimates but the determination of a difference between treatment effects. The preceding discussion of paradoxes is still relevant. First consider its bearing on the test of the parametric null hypothesis of identity of population means.

We have suggested earlier that there are no randomly sampled populations for which experimenters are interested in testing the null hypothesis of identity of means. There is little doubt, though, that some experimenters will persist in basing their statistical conclusions on a model that assumes random sampling of a population, and will draw statistical inferences about *mean* treatment effects. If this is done, the experimenter should be willing to follow through the logical consequence of his decision to draw statistical inferences about means, which is to restrict his statistical inferences about differences between means to the particular measurement scale he employed and to other scales that are linearly related to it. For instance, with circular objects, to reject the null hypothesis of identical mean diameters measured in inches under two treatments is to reject the hypothesis of identical mean diameters measured in centimeters or in millimeters, and to reject the null hypothesis of identical mean circumferences in these units, but it is *not* to reject the null hypothesis of identical mean *areas.* Unfortunately, then, the experimenter starts out with a hypothesis involving a general concept like the size of a circle and is forced, by his decision to deal with inferences about means, to restrict his statistical conclusions to *certain dimensions*.

A randomization test *can* test a hypothesis regarding a general concept. Suppose we randomly assigned subjects to two experimental treatments to determine the relative effect of the treatments on the pupil size. We measure the *pupil diameters,* carry out a randomization test and reject the null hypothesis of no difference in pupil diameter under the two treatments. The alternative hypothesis we accept is that at least one subject would have a different *pupil diameter* under the other treatment. It is necessary then to conclude that at least one subject would have had a different pupil circumference and a different pupil *area,* as well, under the other treatment. To conclude otherwise would result in a contradiction because a change in pupil diameter necessarily means a change in circumference and area, and indeed in *any* measure ordinally related to the diameter. Thus the statistical inference is that at least one subject would have had a different *pupil size,* not just a different *pupil diameter,* under one treatment than under the other.

Assume that we randomly assign subjects to *A* and *B* treatments to find out about the effects of the treatments on speed. We measure the distance run in two minutes and use distance/time as our measure of speed. We predict that the *A* speed will be greater than the *B* speed. We use a randomization test to compare the treatments and reject the null hypothesis in favor of the alternative hypothesis that at least one subject would have run a greater distance under *A* treatment than under *B* treatment. Now any subject who would have run a greater distance under *A* than under *B* during the two minutes would not only have a larger distance/time measurement but would necessarily have a *smaller* time/distance measurement and would have been judged faster in terms of that measure, also. Therefore, when we reject the one-tailed null hypothesis, we accept the alternative hypothesis that at least one subject would have run *faster* under *A* than under *B,* not just that he would have a larger distance/time measurement. Our original theoretical basis for expecting *A* to give the larger distance/time was undoubtedly in terms of the general concept of *speed,* and our conclusion is in terms of the same general concept.

In randomization tests, then, we do not have null hypotheses and alternative hypotheses about highly specific measures but about the broad class of monotonically related measures that define a concept like speed or a concept like the size of a circle. We may compute arithmetic means and differences between means to use as *test statistics,* but our *statistical inferences* refer to individuals and measurements on individuals, not to arithmetic means, and consequently concern more general concepts than would be possible if the inferences were about means. On the other hand, when experimenters test hypotheses about *mean* effects of treatments, their statistical inferences must be restricted to a much narrower class of measures (the class of measures linearly related to the measure employed) than are involved in the concept that led to the experiment.

Normal curve tests as approximations to randomization tests

In the section on estimation it was noted that the assumption of a normal distribution permits much more precise estimation than if no assumptions are made about the population shape. In experiments concerned with random assignment rather than random sampling, however, the normality assumption has no relevance.

The randomization test counterpart to random sampling of a population is random assignment of a set of conditions or individuals. Just as it is impossible to make statistical inferences about population parameters without random sampling, regardless of parametric assumptions, so also is it impossible to make statistical inferences about treatment effects without random assignment, regardless of parametric assumptions.

When an experiment has been designed so that a randomization test can be carried out, a normal curve test can sometimes be used as an approximation to the randomization test. Normal curve tests can be regarded as approximations to randomization tests to the extent that the sampling distribution of the relevant statistic, such as *t* or *F*, underlying the probability tables, is similar to that for a randomization test using the same test statistic.

Consider the *t* test for a difference between means. We deliberately select 10 persons and randomly assign five to treatment *A* and five to treatment *B*, and obtain measurements. There are 252 equally probable divisions of the data in the randomization test, and for each we compute a *t*. We can say that the *t* test is a close approximation to the randomization test if the distribution of *t* for eight degrees of freedom on which the probabilities in the table of *t* are based is approximately the same as the distribution of *t* obtained from the 252 divisions of the data. Of course, this would only show the closeness of the approximation in a given instance. The closeness of the approximation under general conditions has been shown theoretically and by numerical examples by a number of investigators (Silvey, 1954; Wald and Wolfowitz, 1944; Eden and Yates, 1933; Fisher, 1935; Kempthorne, 1952; Pitman, 1937; Welch, 1937).

When normal curve tests are used as approximations to randomization tests, they test the same hypotheses. That is, rejection of the null hypothesis does not imply a difference between means, but a difference between treatment effects on at least one of the subjects. The t or F statistics, although based on a difference between means, are to be regarded as if they were test statistics computed in a randomization test, used to test a hypothesis about treatment effects on individuals.

The randomization-test approach presents a different outlook on testing hypotheses about treatment effects, one which in many ways is perhaps more meaningful to experimenters than the normal curve approach. Therefore, even if a person always uses normal curve tests on his experimental data, he may find it useful to think of them as approximations to randomization tests.

Analysis of variance

In applying analysis of variance in testing for the difference between two treatments when there has been random assignment to treatments instead of random sampling of a population, the tabled .05 and .01 probabilities can be regarded as approximations to the exact probability based on the distribution of F over all possible divisions in a randomization test for independent samples. The exact probability associated with a computed value of F is determined by noting what proportion of the divisions provide such a large value of F.

In fact, it is not necessary to compute F for every division in order to determine the probability value for an obtained F ratio. Instead of computing F, one could compute the absolute difference between means and determine the probability under the null hypothesis of getting such a large absolute difference between means. This probability is exactly the same as the probability of getting such a large F as the obtained F. This will now be shown.

When we divide a single set of measurements in various ways, for a randomization test for independent samples, the degrees of freedom df associated with the sums of squares within treatments SS_w is the same for all divisions, and the degrees of freedom associated with the sums of squares between treatments SS_b also is the same for all divisions. Therefore, the F ratio

$$\frac{SS_b/df_b}{SS_w/df_w}$$

changes from one division to another solely because the ratio of SS_b to SS_w changes. Now the total sums of square $SS_b + SS_w$ is constant over all divisions, because all divisions involve the same numbers, although they are partitioned in different ways. Therefore, when the SS_b increases, the SS_w must decrease, and vice versa. In the F ratio, then, as the numerator increases, the denominator decreases. The SS_b simply reflects the absolute difference between the means, and so as the difference between means increases, the numerator of the F ratio increases and the denominator decreases, thereby increasing the value of F. Thus, the sizes of absolute differences between means among the divisions are in exactly the same order as the sizes of F over the divisions. Consequently, the probability of getting as large an absolute difference between means as the obtained difference is exactly the same as the probability of getting as large an F as the obtained F, in the randomization test for a difference between independent samples.

Perhaps stating the matter differently will help. The SS_b depends on the absolute difference between the means and the SS_w depends on the variability within treatment groups. The means differ most when the lowest scores are in one group and the highest are in the other, making each group maximally homogeneous. The means differ least when each group contains a mixture of high and low scores and therefore is quite variable. Because of this inverse relationship between SS_w and SS_b over divisions of a set of scores, the numerator and denominator of the F ratio vary inversely. Since the numerator varies directly with the absolute difference between the means, the F ratio increases whenever the absolute difference between the means increases.

When there are the *same number* of subjects in both treatments, in the sampling distribution of all equally probable divisions of the measurements there is one division in which treatment A has all the largest scores and another division in which treatment B has all the largest scores. In fact, for every division there is a corresponding division such that the B scores in one division are the A scores in the other. Therefore, for every positive value of $\overline{X}_A - \overline{X}_B$ in the sampling distribution, there is a negative value of $\overline{X}_A - \overline{X}_B$ of the same absolute magnitude. If one uses as a test statistic in a randomization test either the absolute difference between means or F and determines the probability of getting such a large value under the null hypothesis, the probability refers to such a large difference between means in either direction. If the direction of difference is predicted, and if there are *equal numbers* in the two treatments, the probability value for F or for the absolute difference between means can be halved, because of the symmetry of the arithmetic differences between means about the mean difference of 0.

On the other hand, with unequal numbers, there is no justification for halving the probability associated with F or with an absolute difference between means simply because the direction of difference has been predicted. The reason is that the sampling distribution of arithmetic differences between means may not be symmetrical. Suppose we have many small measurements and a few large ones, and that we have more subjects in A than in B. Then, in the sampling distribution of the test statistic, $\bar{X}_A - \bar{X}_B$, the largest *positive* value of $\bar{X}_A - \bar{X}_B$ will be considerably less than the largest *negative* value, because the maximum difference between means will occur when the high and low measurements are maximally separated. This maximal separation of high and low measurements occurs when the few large measurements are in the group with a small number of subjects (treatment-B group) and the many small measurements are in the group with a large number of subjects (treatment-A group) as there is then the least mixing of large and small measurements within groups. In such a case, if we computed the probability of such a large *absolute* difference (or F) as we obtained, and determined this to be .10, we should not divide this value by 2 to get a one-tailed probability of .05, simply because we correctly predicted $\bar{X}_B > \bar{X}_A$ because this exaggerates the significance of the result. The actual probability of getting such an extremely large *negative* value of $\bar{X}_A - \bar{X}_B$ as the obtained value is likely to be somewhere between .05 and .10, because of the negative values in the sampling distribution of $\bar{X}_A - \bar{X}_B$ being larger than the positive values.

The preceding discussion concerns matters that have occasionally been considered in regard to the random sampling model of analysis of variance: the effect of skewness on one-tailed and two-tailed tests and the effect of unequal sample sizes.

t test for paired comparisons

A formula for *t* used in testing for the difference between means of paired measurements is

$$t = \frac{\Sigma D}{\sqrt{[N\Sigma D^2 - (\Sigma D)^2]/(N-1)}}$$

where ΣD is the sum of the differences between paired measurements, that is, $\Sigma(A - B)$. What if we were to use the randomization test for paired comparisons to determine the exact probability for a *t* as large as an obtained *t*, for a situation where there has been random assignment instead of random sampling? We would take our *N* pairs of measurements and determine every arrangement that can be obtained by reversing measurements within pairs. There are 2^N such arrangements. For each arrangement we would compute a value of *t*. The proportion of arrangements with a t as large as our obtained t is the probability of getting a t as large as ours under the null hypothesis. Probability values given in tables of t can be regarded as approximations to the exact probability obtained by the randomization test for paired comparisons, using t as our test statistic.

Actually, one need not go through all of the computation of t for the randomization test. Notice that in the formula for t, N will be the same for all 2^N arrangements because we use the same pairs of measurements (and therefore the same *number* of pairs of measurements) in all arrangements. Also, ΣD^2 is the same for all arrangements because the D^2 value associated with each pair of measurements is the same for all arrangements; only the *sign* of D changes, and this does not affect D^2. Therefore, ΣD and $(\Sigma D)^2$ are the only terms in the formula that change from one arrangement to another in the sampling distribution. ΣD is in the numerator, and ΣD^2, which is in the denominator, is *subtracted* from $N\Sigma D^2$. As ΣD increases, therefore, the numerator increases and the denominator decreases. Therefore, the larger the ΣD, the larger the value of t for an arrangement. Since t and ΣD vary directly over all arrangements, the probability of getting a t as large as an obtained t for paired comparisons is exactly the same as the probability of getting a ΣD as large as an obtained ΣD. ΣD is the statistic used in the section "Randomization test for paired comparisons."

The test statistic ΣD is used for the same reason that it is used in the regular paired comparison t test, namely, because it makes use of the size of the difference within paired measurements. Reversals of measurements where there is a small difference are given less weight because small differences can be expected to occur even without a difference in treatment effects solely from variability within a subject.

If we predict that $\Sigma(A - B)$, in other words ΣD, will be positive, a large difference in a single subject in the opposite direction could keep us from getting significance because this could reduce $\Sigma(A - B)$ as much as would four or five reversals of direction for smaller differences, giving a large number of $\Sigma(A - B)$ test statistics in the sampling distribution as large as the obtained $\Sigma(A - B)$ and thereby making the probability large.

Now suppose that subjects with large measurements were more variable than subjects with small measurements. In such a case we would consider that a large reversal for large measurements was as likely to happen from intraindividual variation as a small reversal for small measurements and would want our test statistic to be affected by the size of the difference between the paired measurements, *relative to the size of the measurements*. For this purpose one could carry out a

166 Statistical Inference: The Distribution-free Approach

logarithmic transformation on the original measurements and compute ΣD as the test statistic for the logarithms.

Product-moment correlation

A formula for the product-moment correlation coefficient that is convenient for machine computation is this:

$$r = \frac{N\Sigma XY - \Sigma X \Sigma Y}{\sqrt{[N\Sigma X^2 - (\Sigma X)^2][N\Sigma Y^2 - (\Sigma Y)^2]}}$$

Suppose we were to use a randomization test for correlation to determine the exact probability for a product-moment correlation coefficient where we have random assignment instead of random sampling. We first compute r for the obtained pairing of measurements. Then we would pair our measurements in every possible way and compute r for each pairing. The probability of getting such a large r as our obtained r would be the proportion of r's among all pairings that were as large as the obtained r.

It is not necessary to do so much computation, however, for the randomization test. Notice that in the above formula, every term in the denominator will be the same for all pairings, and that $\Sigma X \Sigma Y$ also is constant over all pairings. And, of course, so is N. Therefore, the only thing that produces variation in the value of r over all pairings is ΣXY. The larger the value of ΣXY the larger the value of r. Thus the probability of getting an r as large as an obtained value is exactly the same as the probability of getting as large a ΣXY as the obtained ΣXY. (ΣXY is the statistic used in the section "Randomization test for correlation.")

So far we have not referred to the fact that the probabilities associated with the various test statistics we have proposed for randomization tests are constant (stay the same) under linear transformations of the measurement numbers. It is not hard to demonstrate this kind of constancy for any of the tests. It will be shown here with regard to the correlation statistic ΣXY.

Let $aX + b$ refer to a linear transformation of X and $cY + d$ refer to a linear transformation of Y. Then the product of a pair of transformed X and Y measurements is

$$(aX + b)(cY + d) = acXY + bcY + adX + bd$$

and the sum of the products of the transformed measurements is

$$ac\Sigma XY + bc\Sigma Y + ad\Sigma X + Nbd$$

We can ignore those terms in the expression which are constant over all pairings in the sampling distribution because they do not affect the probability value associated with the test statistic. All the terms are constant except ΣXY; therefore, the sum of the products of linearly transformed X and Y measurements varies directly with the size of ΣXY, and so the computed probability values are the same for the transformed and the untransformed measurements.

Must null hypotheses be false?

A number of persons believe that the null hypothesis of no difference in treatment effects is almost certain to be false for psychological experiments. We will consider two arguments that are given.

One argument refers to the inappropriateness of the null hypothesis of identical mean effects of *treatments* but is based on consideration of differences between means of *existing populations* of people. It is argued that almost never will two different groups of individuals have exactly the same mean age, the same mean reaction time, or the same mean value of any other attribute.

This argument is founded on a misunderstanding. In experiments based on random assignment to treatments, the population means which the (parametric) null hypothesis specifies to be equal are not the means of two different groups of individuals but of the same individuals under two treatments. We do not in experimental situations ordinarily randomly sample two different groups of people. If we want to refer to the random assignment procedure as a random sampling procedure, we can say that we take a random sample from a single "population" of individuals (our experimental subjects) for which there are two hypothetical distributions. If the *distributions* are called "populations," we have two populations of measurements from the same individuals. Two such "populations" will have identical measurements (and consequently identical means) whenever the treatment effects are identical.

The other argument is less easy to dismiss. The argument is simply that in the typical psychological experiment, people in general, including the experimenter himself, know that the independent variable will have some effect on the responses, and so the null hypothesis of identical treatment effects or identical mean effects can usually be rejected without a test. Persons advancing this argument recommend that some other procedure, such as an estimation procedure, be used instead of hypothesis testing.

The argument seems reasonable enough with regard to the typical psychological experiment. Probably hypothesis-testing procedures are not the best statistical procedures for analyzing experimental results in most cases. Unfortunately, the proponents of this argument are inclined to act as though null hypotheses for psychological experiments are *never* plausible and to make a blanket recommendation that hypothesis-testing procedures be replaced by other statistical procedures. This recommendation should not be taken seriously because there certainly are null hypotheses for psychologists to test which are plausible and yet not so certain as to make a test unnecessary.

It is not hard to imagine realistic experiments in which the null hypothesis of identical treatment effects would be considered by many persons to be plausible and definitely worth testing. For example, consider the null hypothesis that a blindfolded person who claims ability to discriminate colors tactually will respond the same at any stimulus presentation, whatever the color of the paper he touches. Or the null hypothesis that a person will not be influenced by a suggestion projected on a screen for such a brief time that he is unaware that he has read anything. Or consider the null hypothesis that the cortical activity in a given region of the brain will not be affected by stimulation of another specified region of the brain within one-tenth of a second. Such a null hypothesis might seem plausible to those persons who assumed that certain neural pathways would be used and who assumed certain rates of neural transmission.

In spite of the rareness of psychological experiments where the null hypothesis of no difference between treatment effects is a plausible null hypothesis, when a person *does* have such an experiment and can reject the null hypothesis the experiment is likely to be judged important, because the experimenter has found something contrary to what many people expected. Tests of null hypotheses are more appropriate for experiments that explore new areas than for those that are intended to polish up well-known facts.

Chapter Six

Experimental Regression

In the preceding chapter it was indicated that actual existing populations are not those that the experimenter wants to draw inferences about; that the experimenter usually studies a particular group of individuals only for the purpose of drawing *statistical* inferences about treatment effects on them in order to generalize on a *nonstatistical* basis to individuals of the same kind. The study of psychophysical and other functional relationships in psychological research will be examined here from the same standpoint; the population of interest is assumed to be one that cannot be randomly sampled, and the justification for statistical inference is random assignment. Taking this approach rules out the use of the usual regression procedures for drawing inferences for the same reason that they were earlier ruled out (in discussions of estimation) for making inferences about populations in the future and other populations that cannot be randomly sampled. Again we will consider both the statistical inferences that can be made and the conditions that facilitate nonstatistical generalization, that is, the making of nonstatistical inferences.

The terms *experimental regression* and *nonexperimental regression* will be used to refer to regression analysis in experimental and nonexperimental research, respectively.

The objectives of experimental regression

The objectives of experimental regression differ from those of nonexperimental regression in two major respects. One of these is the type of inference that is made and the other is the type of population about which the inferences are made.

Now nonexperimental regression may lead to the inference that one variable is correlated with another within a population, but such regression provides no basis for statistical inferences about causal relations. Experimental regression, on the other hand, being based on research where an independent variable is manipulated, permits causational inferences about the effect that variation in the independent variable has on the dependent variable—or at least the effect that a particular way of varying the independent variable has on the dependent variable.

The type of population with which nonexperimental regression is concerned is an existing population. It is frequently defined in geographical, political, or social terms. Experimental regression, on the other hand, is concerned with nonexistent populations, which are populations of individuals (including individuals from the past and future) defined in terms of the characteristics they possess independently of group membership. Experimental regression has the objective of making inferences that apply to every individual of a particular kind rather than inferences about average measurements of a heterogeneous group of individuals.

Experimental-regression techniques are used to determine something more than simply whether variation in an independent variable produces variation in a dependent variable. There is also the aim of finding out something about the general nature of the relationship between the two variables, such as whether it is linear or exponential. And in a number of instances, it is of scientific importance to psychologists to have estimates of numerical values of the dependent variable that can be expected from setting the independent variable at certain values.

General laws

In comparisons of two treatment groups it is usually recognized that the conclusions must be restricted to situations comparable to the experimental situation because a number of factors in the experimental situation may influence the experimental results. There is little temptation to say that treatment A has more effect than treatment B without specifying some of the assumed environmental limiting factors. On the other hand, experimental regression is sometimes used to look for the "true" relationship that holds between two variables, without proper consideration for the dependence of the observed relationship on other factors than the designated independent variable. But conclusions about the relationship between two variables can be no more general than conclusions about differences between two treatments.

For example, data from a single experiment cannot yield a curve that shows how manipulation of intensity of light affects constriction or dilation of the pupil of the eye, except for a particular set of conditions. There are several conditions that can affect the curve. The curve will be affected by certain drugs that the person may take. Also, if the light is flickering instead of steady, the effect on the pupil is different. Furthermore, whether the light intensity is increased or whether it is decreased affects the curve. Whether the person is light-adapted or dark-adapted at the time also would have some influence. And there would be a difference between the effect of a gradual change in intensity and an abrupt change. And, of course, the curve for the response of the pupil to changes in intensity for extremely short exposure times would be different from the curve for longer exposure times.

The demands of application may cause us to focus on "realistic" situations. Industrial organizations, for example, might be interested in the response to long exposure times, with white light, for persons who do not take drugs that affect pupil constriction, etc. Practical uses of experimental findings are important, but we must realize that experiments that deal with healthy, strongly motivated persons with good sensory and motor systems, in experimental situations simulating everyday conditions, do not provide curves showing the "true" relationship between two variables. The relationship found is a specific relationship that may have some predictive utility because the specific conditions are commonplace, but the relationship is not the *true* relationship just because it "generally holds." A more useful attitude would be that the relationship found holds only for the type of situation under which it was obtained and that such situations are frequently found. A regression line may be useful because the conditions under which it was obtained are frequently encountered, but one should not overlook the importance of other regression lines for less common situations.

Aspects of design to facilitate nonstatistical generalization

Since the objective of experimental regression is to allow predictions not about a randomly sampled population but about individuals with certain characteristics, the sample used in the experiment should *not* be randomly selected. Random selection for experimental regression could provide a very heterogeneous sample, making it difficult both to detect clear-cut relationships and to know to what kind of individuals the results can be generalized. The problems are precisely those of the types of experiments that were discussed in the preceding chapter and are solved in the same way. The sampling procedure should follow the lines laid out for comparisons of experimental treatments. Where possible, all the experimental measurements should be obtained from one individual. Where it is not possible to get enough measurements from one subject, other subjects that are selected should be homogeneous with regard to characteristics thought to be relevant to the treatment effects. After a regression line has been determined for an individual, the uniqueness of the characteristics of the individual make it difficult to know which characteristics are relevant to the experimental results and which are not relevant. To investigate this, we can replicate the experimental study with other individuals, or groups of individuals where more than one individual is necessary, selected to differ with regard to a single variable, such as sex, age, or weight to determine the relevance of this variable.

Fitting regression lines to data

A key problem in regression analysis is to select a method of fitting regression lines that will provide lines of best fit. There are various criteria that could be used to determine which of several lines of a given type best fit a set of data, but the least-squares criterion is commonly employed. For example, in linear regression, the line of best fit for predicting Y from X is said to be the straight line for which the sum of the squared vertical deviations of the measurements from the line is a minimum. Since the variance is the average squared deviation, the criterion is effectively one of requiring the minimum variance about the regression line.

But if a line were simply to be a line of best fit to the data and to be determined entirely by the data, why use a straight line? Would not a wiggly line that passes through every plotted point fit the data better? It is quite clear that nobody fits a line to his data on the basis of the data alone. The fitting of a regression line is based on both the obtained data and theoretical considerations.

We do not fit just any type of line to the data, but a straight line or an exponential or some other specific type of line. What considerations are involved in selecting the type of line to fit to a set of data? If we have a reasonably large sample, the sample relationship itself may suggest the general nature of the relationship. But theoretical considerations also are important.

A linear relationship, for example, implies that equal changes in X produce equal changes in Y. This is a reasonable assumption over certain ranges of X only, and so there is always the problem of judging whether extrapolation (or in some cases even interpolation) should be linear. And if we deal with extremely small or extremely large values of X, we may reach a point where the effect on Y has reached its minimum or maximum value, so that variation of X beyond those points provides no change at all in Y.

Sometimes there are theoretical reasons for expecting an exponential relationship. For example, we may expect X and Y to be related as the area of a circle is related to the radius, $A = \pi r^2$. This kind of relationship is indicated by a straight line on log-log paper because $\log A = \log \pi + 2 \log r$.

Frequently there is reason to expect a regression line to pass through or close to zero at the lower left-hand corner of the graph. For instance, we would expect estimates of weights to come closer and closer to zero with reduction in the physical weight being estimated. There are statistical techniques for constructing lines of best fit that pass through zero. When there is good reason to expect the line to pass through zero, this constraint should be placed on the line of best fit.

Spacing of the independent variable

In nonexperimental regression based on random sampling of existing populations, the number of Y measurements associated with any given X value is not under the control of the experimenter but is determined by the sample.

In experimental regression, on the other hand, we can select any values of X we please and control the number of observations made at each of those values. Suppose we are rather certain that the relationship between two variables is linear. To get a good estimate of the line, we could take repeated measurements at two extreme values of X. The farther apart the two points on the X axis, the more definite the straight line to be drawn through the two sets of points. In using two points, we are assuming that we do not have reason to believe that the line will pass through the zero point or any other specified point, or else we would need to make estimates at only one point. Therefore, each of the two points at which we take measurements is equally important in determining the straight line, and there is no object in making a set of measurements at one point more accurate at the expense of the accuracy of the other set. If one set of measurements is more variable than the other, it is desirable to take relatively more readings at the more variable position in order to bring the accuracy of the mean of this set up to the accuracy of the other set.

Estimating the simultaneous-measurement regression line

In nonexperimental regression, the regression lines for samples can be regarded as estimates of population regression lines that are lines of best fit for a bivariate distribution of scores. There are two regression lines, one which is best for predicting Y from X and one for predicting X from Y. On the other hand, there is only one regression line for experimental regression. We manipulate variable X and take measurements on Y, and so we can predict from X to Y only. In fact, it would usually not make sense to ask what value of the *independent* variable would result from manipulating the *dependent* variable.

Our experimental regression line is an estimate of an ideal regression line. That is, it is an estimate of a regression line which is physically unattainable, although conceptually meaningful inasmuch as we can describe ways to approach the ideal. There are various concepts of the ideal that is to be estimated, depending on the objectives of the experimenter and the types of estimation his experimental procedure permits.

One type of ideal regression line could be called the *simultaneous-measurement regression line* because it is a line conceived as passing through simultaneous measurements on an individual at various magnitudes. (*Simultaneous* measurements would eliminate the possibility of variation in the measurements over time or of an early act of measurement affecting a later measurement.) Or we could say that the simultaneous-measurement regression line refers to a line passing through the various alternative measurements that a subject might have given at a specified time. For example, if we randomly assign a subject to one stimulus magnitude at a specified time, we can imagine that he would have given measurements for other values of the X variable if he had been assigned to them at that time. It is meaningful, therefore, to ask what response he might have provided if he had been assigned a different stimulus magnitude. The simultaneous-measurement regression line would be a line that passed through each of these potential response values, and through the obtained response value, there being one response value for each stimulus magnitude that could have been assigned.

We have used the concept of simultaneous-measurement ideal regression line, without labeling it as such, in our discussion of the randomization test for correlation. The ideal line was postulated by the null hypothesis, not estimated from obtained data, however. The null hypothesis of identical effects for all values of X at any time can be restated as a null hypothesis asserting that the simultaneous-measurement ideal regression line for the subject at a specified time is a straight line parallel to the X axis.

Without complicating the concept of the simultaneous-measurement ideal regression line, we can admit to a slight variation in the ideal lines over the span of time during which we obtain measurements from our subject. Assuming the variation to be slight, we can regard a single regression line as an estimate of *any* of the ideal lines for the times during which the measurements were made.

Estimating the perfect-experimental-control regression line

Procedures for generalizing from an obtained regression line to one that we might expect under ideal experimental conditions are not readily available because the practice has not been explicitly carried out. Although experimenters do not state that they are generalizing to situations with better or perfect experimental control, there is little doubt that this is what many of them have in mind in working up a regression line. These experimenters use their obtained regression lines to estimate what might be called the *perfect-experimental-control regression line.* Whether or not the notion of an ideal regression line as such crosses their mind, their conclusions regarding the relationships found in the experimental situation are discussed as though they were not confined to the particular degree of experimental control employed. Certainly the experimenter would expect and hope that with better experimental control the regression line would be like the one he has. It may seem to some people that in fact experimenters do not use the concept of a perfect-experimental-control regression line, explicitly or implicitly, but simply consider generalizing to *better-controlled* experiments. This appears not to be the case, since they do not act as though there is any limit in the degree of control beyond which they would not expect a closer approach to their regression line. This implies that they regard their regression line as an estimate of one which is approached without limit as one improves the control over experimental variables.

Now we can never eliminate the adverse effect of stray noises, lights, magnetic interference, and other undesired distractions on our experimental outcomes, but this does not mean that the concept of the ideal perfect-experimental-control regression line is not a useful one. Even though in real life we can never reach the ideal but can only approach it, there is value in having a concept to guide our thinking. We shall, however, see in the following discussion of symmetry about a regression line (Edgington, 1966b) how difficult it is to define the situation which is to produce this ideal regression line and to justify extrapolation from experiments that fall short of the ideal situation.

Implications of symmetry about a regression line

Suppose we performed a psychophysical experiment in which inadequate soundproofing permitted auditory distractions to contribute to the variability of the responses. How can we know whether the regression line for predicting response values from stimulus values should be regarded as a line that shows the response values which would be approached under better experimental control (in the psychophysical experiment, better control over the auditory distractions)? We will not attempt to provide a comprehensive answer but will attempt to evaluate the importance of one factor: the symmetry of responses about the regression line.

If the distribution of responses appears to be symmetrical about the regression line, with the measurements clustering near the line and becoming widely scattered away from it, the regression line looks like an estimate of the values that would be approached by the responses as experimental control improved. When, on the other hand, the distribution of responses about a regression line is skewed, the regression line seems more arbitrary. In some cases, however, a transformation of the response measurements in a skewed distribution will provide a symmetrical distribution of the transformed measurements about the regression line. For example, for positively skewed distributions of responses, a logarithmic transformation may result in a symmetrical distribution of transformed measurements. The regression line of the raw measurements is a line of best fit to the response means. The regression line for the transformed measurements is a line of best fit to the logarithms of the geometric means of the response measurements, since the regression line is fitted to the means of the logarithms of the response measurements. To assume that reduction of the variability of the responses causes the distribution of the *logarithms* of the response measurements to move in symmetrically toward the regression line is to assume that the *raw measurements* would move toward the regression line by an amount proportional to the distance of the measurements from the regression line.

When responses are symmetrical about the regression line, it appears intuitively plausible that the measurements would move symmetrically inward toward the regression line if the variability of the responses were reduced, but we should examine the possible bases of this assumption to see if it is justified. Consider three arguments in favor of this assumption. (1) If we assume that certain "random" processes led to the obtained symmetrical dispersion of the response measurements about the regression line, elimination of those processes should provide a symmetrical *reduction* of the dispersion. (2) Another argument is that, if the variability is to be reduced, we can expect the measurements to move in from one side of the regression line in the same way as they do from the other side, because there is no good reason to expect otherwise. (3) A third argument is that it is unlikely that the symmetry of the obtained distribution is unique to the particular degree of experimental control used, and so increased experimental control should maintain the symmetry while reducing the variability.

A hidden assumption in each of these arguments is that better experimental control reduces variability but does not raise or lower the general level of the response measurements. This assumption is not always tenable. Variability is eliminated by holding a variable constant, but it is held constant at some particular level. When the values of the variable being held constant are correlated with the response values, the particular level at which the variable is held constant will determine the general level of the responses. That is, not only will the variability of the responses be reduced, but the level of the response measurements also may shift. For example, in studies of certain types of learning, efforts to reduce the variability of motivation may involve trying to hold motivation constant by maximizing it. Better control over variability of motivation would then not only reduce the variability of learning measurements, but would shift the distribution upward, as well.

There are special situations, however, where shifting the level at which a variable is held constant can reduce variability without shifting the response distribution. For example, lowering the level of an auditory distraction by improving soundproofing might reduce the variable error of visual estimates of line length without changing the level of the estimates because the variable to be controlled is not correlated with the response values but is correlated with response variability. In such cases, if the responses are symmetrically distributed about the regression line, it is justifiable to treat the regression line as the line which is approached by the responses as response variability is reduced.

Randomization test for a difference between regression lines

Suppose we want to compare two regression lines for the same values of an independent variable under two treatment conditions. Furthermore, assume that each subject can provide only one measurement. We select 10 subjects, as similar as possible with regard to characteristics we think are relevant. Then we split our subjects into five pairs of subjects, striving for maximum homogeneity within pairs. Each pair of subjects is assigned one of five stimulus magnitudes, and

178 Statistical Inference: The Distribution-free Approach

each subject in each pair is assigned a treatment time. Then one member of each pair is *randomly* assigned to treatment A, the other being assigned to B. The measurements are taken for the previously assigned stimulus magnitudes and treatment times. The null hypothesis is that each measurement is the same as it would have been under the other treatment. We have 2^5 equally probable divisions of our 10 subjects, involving all possible ways of dividing the five pairs of subjects so that one member is assigned to A and the other to B. Under the null hypothesis each person gives exactly the same response to the stimulus as he would have if he had received the stimulus at that time under the other treatment condition. Rejection of the null hypothesis implies that at least one of the subjects would have responded differently to the same stimulus magnitude under the other treatment.

The test statistic to be used depends on the type of difference expected. For example, if we expected a difference between the levels of the two sets of responses, the test statistic could be the difference between the means of treatments A and B. On the other hand, if we expected a difference in the *slopes* of the regression lines, but not necessarily in the overall level, we would use a different test statistic. One such statistic would be the ratio of the slopes of the regression lines for the two treatments. For a one-tailed test where we predict that treatment A gives a steeper slope, we fit linear regression lines to the two sets of data for each of the 2^5 equally probable arrangements, and for each of the arrangements, use the ratio

$$\frac{\text{Slope of treatment-}A\text{ regression line}}{\text{Slope of treatment-}B\text{ regression line}}$$

as the test statistic.

A statistical test for nonmonotonic trends

When a functional relationship departs considerably from linearity, a product-moment correlation coefficient may not be significant in a situation where a rank-order correlation coefficient would be. Thus, we can regard the product-moment correlation technique as being useful for detecting linear relationships, and rank-order correlation techniques as being useful for detecting monotonic relationships of various sorts, linear or nonlinear. But what about the detection of nonmonotonic relationships, such as those in which manipulation of the independent variable produces a gradual rise in the magnitude of the dependent variable, followed by a gradual fall, or vice versa? In other words, what about detecting U-shaped or inverted U-shaped relationships? A technique sensitive to nonmonotonic as well as monotonic trends would be appropriate. Such a technique has been developed and a probability table for it has been published (Edgington, 1961a). The following discussion of the application in testing Weber's law will show how to use the technique.

The study to be described (Edgington, 1961b) was conducted for the purpose of testing Weber's law with data that had been widely published. Since it is generally agreed that Weber's law does not hold for extreme intensities, only data for the middle range of stimulus-intensity were analyzed statistically. Visual brightness discrimination data, namely, Konig's data for white light (Blanchard, 1918), were used. These data were particularly appropriate because they had been used by some investigators for showing the *tenability* of Weber's law and by other investigators for showing the *untenability* of the law.

The statistical test applied to the data utilizes quantitatively the aspects of the data that investigators have used qualitatively in rejecting Weber's law: the gradual fall and gradual rise of the Weber-ratio curve (Boring, 1942; Holway and Pratt, 1936).

Table 6.1 Visual-brightness discrimination for white light (data in millilamberts)

Log intensity	−1.70	−1.40	−1.10	−0.70	−0.40	−0.10	0.30	0.60
Weber ratio	.0560	.0455	.0380	.0314	.0290	.0217	.0188	.0175
Sign of difference	−	−	−	−	−	−	−	+

Log intensity	0.90	1.30	1.60	1.90	2.30	2.60	2.90	3.30	3.60
Weber ratio	.0178	.0176	.0173	.0172	.0170	.0191	.0260	.0266	.0346
Sign of difference	−	−	−	−	+	+	+	+	

SOURCE: Edgington, E. S. 1961. A statistical test for cyclical trends, with application to Weber's law. *The American Journal of Psychology,* **74** (no. 4), 630-632.

The signs shown in Table 6.1 indicate an increase (+) or decrease (−) in the size of the Weber ratios as the stimulus-intensity increases. The test statistic is the number of runs of signs of successive differences, where a run is a group of *one or more* identical signs. When numerical values are arranged randomly in a series, one can expect considerable fluctuation and therefore several runs. On the other hand, when numerical values vary systematically in a series, one usually expects few runs. The set of 17 Weber ratios shown in Table 6.1 contains four runs: a run of seven minuses, followed by a run of one plus, then a run of four minuses, and finally a run of four pluses. Table 6.2 is the probability table for the test. The intersection of row 4 (for the number of runs) and column 17 (for the number of observations) shows a value of .0000. This indicates that the probability of getting as few as four runs with 17 randomly arranged observations is less than .0001. The conclusion is that Weber's law did not hold in the experiment providing this set of data.

Some persons would argue that tests of a null hypothesis have no relevance to Weber's law. They would say that Weber's law is not assumed to be an exact law but is intended to be a practical approximation. Weber's law holds, they would say, in the case of the brightness-discrimination data, because the Weber ratios are *almost* constant over the range of stimulus intensities. Any of a number of quantitative laws would be said by these people to hold for data on sensory discrimination because they would provide approximations adequate for practical use.

There are other persons, however, who are concerned with theoretical considerations, and they would argue that any evidence of a *systematic* departure from constancy of the Weber ratio over a range of stimulus magnitudes, no matter how slight the departure may be, is of theoretical importance. These people would be interested in testing a null hypothesis regarding Weber's law. Presumably to them Weber's law would imply something like the existence of a simultaneous-measurement ideal regression line for the Weber ratio that is horizontal to the base line.

Now in fact, although we analyzed data to demonstrate the use of a statistical test, the data were not gathered by us, and so we do not know whether the conditions of random assignment justified its application. We will, however, consider how the preceding data should have been gathered to permit the use of this statistical test.

First, we select one subject. We decide on 17 treatment times. Then we randomly draw the treatment times to associate them with 17 predetermined light intensities. Our null hypothesis is that the Weber ratio (just noticeable difference in intensity divided by the intensity) for any given treatment time is exactly what it would have been under any of the other intensities. With our random assignment of treatment times, under our null hypothesis there are 17! equally probable pairings of the obtained Weber ratios with the 17 light intensities. If we computed the number of runs of signs for each pairing, what proportion of the 17! arrangements would have as few as four runs of signs? Less than .0001. Therefore, we reject the null hypothesis and conclude that for at least one of the treatment times the subject would have provided a different Weber ratio at one of the other stimulus intensities.

Table 6.2 Probability of r or fewer runs of signs of first differences for o observations

Number of runs (r)	Number of observations (o)																								
	1	2	3	4	5	6	7	8	9	10	11	12	13	14	15	16	17	18	19	20	21	22	23	24	25
1		1.0000	.3333	.0833	.0167	.0028	.0004	.0000	.0000	.0000	.0000	.0000	.0000	.0000	.0000	.0000	.0000	.0000	.0000	.0000	.0000	.0000	.0000	.0000	.0000
2			1.0000	.5833	.2500	.0861	.0250	.0063	.0014	.0003	.0001	.0000	.0000	.0000	.0000	.0000	.0000	.0000	.0000	.0000	.0000	.0000	.0000	.0000	.0000
3				1.0000	.7333	.4139	.1909	.0749	.0257	.0079	.0022	.0005	.0001	.0000	.0000	.0000	.0000	.0000	.0000	.0000	.0000	.0000	.0000	.0000	.0000
4					1.0000	.8306	.5583	.3124	.1500	.0633	.0239	.0082	.0026	.0007	.0002	.0001	.0000	.0000	.0000	.0000	.0000	.0000	.0000	.0000	.0000
5						1.0000	.8921	.6750	.4347	.2427	.1196	.0529	.0213	.0079	.0027	.0009	.0003	.0001	.0000	.0000	.0000	.0000	.0000	.0000	.0000
6							1.0000	.9313	.7653	.5476	.3438	.1918	.0964	.0441	.0186	.0072	.0026	.0009	.0003	.0001	.0000	.0000	.0000	.0000	.0000
7								1.0000	.9563	.8329	.6460	.4453	.2749	.1534	.0782	.0367	.0160	.0065	.0025	.0009	.0003	.0001	.0000	.0000	.0000
8									1.0000	.9722	.8823	.7280	.5413	.3633	.2216	.1238	.0638	.0306	.0137	.0058	.0023	.0009	.0003	.0001	.0001
9										1.0000	.9823	.9179	.7942	.6278	.4520	.2975	.1799	.1006	.0523	.0255	.0117	.0050	.0021	.0008	.0003
10											1.0000	.9887	.9432	.8464	.7030	.5369	.3770	.2443	.1467	.0821	.0431	.0213	.0099	.0044	.0018
11												1.0000	.9928	.9609	.8866	.7665	.6150	.4568	.3144	.2012	.1202	.0674	.0356	.0177	.0084
12													1.0000	.9954	.9733	.9172	.8188	.6848	.5337	.3873	.2622	.1661	.0988	.0554	.0294
13														1.0000	.9971	.9818	.9400	.8611	.7454	.6055	.4603	.3276	.2188	.1374	.0815
14															1.0000	.9981	.9877	.9569	.8945	.7969	.6707	.5312	.3953	.2768	.1827
15																1.0000	.9988	.9917	.9692	.9207	.8398	.7286	.5980	.4631	.3384
16																	1.0000	.9992	.9944	.9782	.9409	.8749	.7789	.6595	.5292
17																		1.0000	.9995	.9962	.9782	.9563	.9032	.8217	.7148
18																			1.0000	.9997	.9846	.9892	.9679	.9258	.8577
19																				1.0000	.9975	.9983	.9924	.9765	.9436
20																					.9998	.9999	.9989	.9947	.9830
21																					1.0000	1.0000	.9999	.9993	.9963
22																							1.0000	1.0000	.9995
23																									1.0000
24																									1.0000

Let us next consider how to apply this test when there are tied measurements. When the tied measurements are not adjacent there is no need for adjustment. But, suppose the value .0290 had been .0314 in Table 6.1, so that there were two adjacent tied values. In that case, we remove either one of the .0314 values from the series and carry out the same test as before with only 16 observations instead of 17.

The sampling distribution of numbers of runs for $N > 25$ is very close to a normal distribution. The mean number of runs is

$$\frac{2N - 1}{3}$$

and the number of runs is distributed almost normally about this mean with a standard deviation of

$$\sqrt{\frac{16N - 29}{90}}$$

where N is the number of observations. The significance of the number of runs obtained can be determined from normal curve tables. To correct for the discontinuity or discreteness of the distribution of number of runs, it is necessary to reduce the obtained deviation from the mean by .5.

Chapter Seven

Joint Analysis of Differences in Central Tendency and Variability

In the previous two chapters we considered the application of randomization tests to a variety of experimental situations. In this last chapter we will concentrate on a single complex problem: how to draw inferences from joint analysis of differences in central tendency and variability. At first we will deal with the problem without regard to randomization tests. Then the problem will be re-examined from the standpoint of randomization tests to see what light the distribution-free approach casts on the problem, and to demonstrate potentialities of randomization tests that were not apparent in the simpler applications.

Comparisons of frequency distributions

The term *differential effect* will be used to refer to the difference between the magnitude of a subject's *obtained* measurement and the measurement he *would have obtained* under the other treatment. To illustrate, the two-tailed null hypothesis for many randomization tests can be stated in this form: The differential effect for each subject is zero.

Figure 7.1 shows a pair of frequency distributions for which the differential effect (treatment-*B* measurement minus treatment-*A* measurement) for each subject is three points. The identity of the subjects in the two distributions is shown by the code letters. The frequency distributions differ only by a translation of three points; their shapes are the same. Figure 7.2 shows a frequency-distribution difference associated with a correlation between the differential effect and the measurement magnitude. In this case the differential effect is one-half the measurement under treatment *A*. Since the low subjects have smaller differential effects than the high subjects, the frequency distribution with the larger mean is more variable.

184 Statistical Inference: The Distribution-free Approach

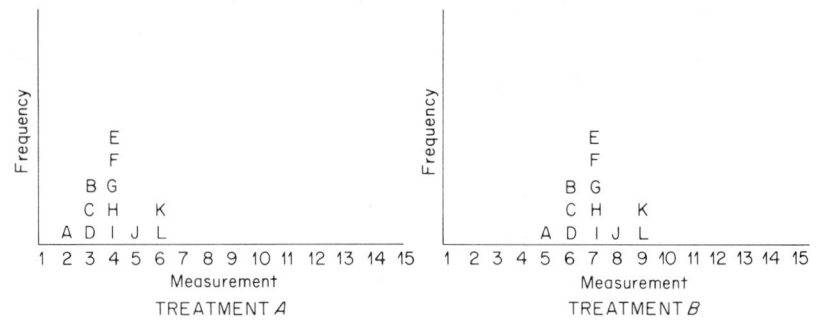

Figure 7.1 Distributions illustrating identical differential effects for all subjects.

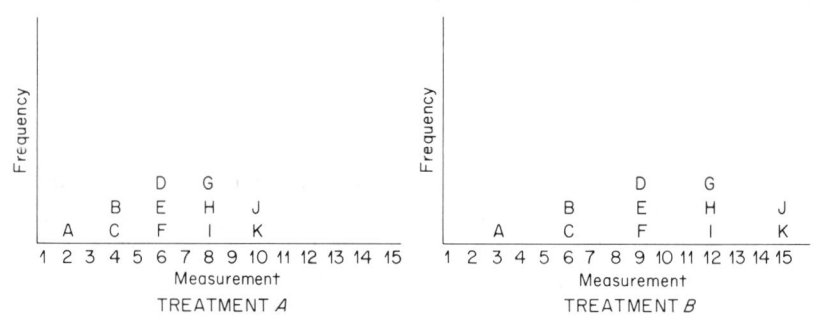

Figure 7.2 Distributions illustrating differential effects that are positively correlated with the measurement magnitude.

Types of differences in central tendency and variability

In the following analysis it will be assumed that there is a high positive correlation between the measurements of persons under one treatment and what the measurements of the same persons would have been under the other treatment. More specifically, we will assume that persons at the low end of the distribution of measurements for one treatment would have been at the low end of the distribution for the other treatment if they had been subjected to it, and that persons at the high end of the distribution would have been at the high end of the distribution under the other treatment. This is a plausible assumption since the same attribute, for example, reaction time, strength, weight, or metabolism, is measured under both treatments.

Consider a hypothetical situation involving two groups of individuals. One group is taught by method *A* and the other by method *B*. On an academic achievement test, method *B* yields the larger mean score and the smaller variability. The following diagram represents the two distributions of achievement-test scores, the length of the line indicating the variability and the position of the line the size of the scores:

Method *A*: ─────────────
Method *B*:   ─────────

The relationship of the lines to each other indicates the overlap of the distributions. There is a greater difference between the low ends of the distributions than between the high ends, suggesting that the differential effect is greater for low-scoring persons than for high-scoring persons.

It is conceivable that the test scores associated with method *B* were so high that the maximum possible test score restricted the variability, but whether this was the case could not be known without additional data. The diagram of the distribution overlap simply suggests this possibility.

The particular combination of direction of difference in central tendency and direction of difference in variability in the above example is one of five combinations that will be analyzed to compare differences between distributions at the low ends with differences at the high ends. Examples will be given to show the value of such analysis in interpreting research data and suggesting further research.

The classification system shown in Table 7.1 (Edgington, 1963) classifies situations with distributions from two treatments ac-

Table 7.1 Classification of differences in central tendency and variability

Situation designation	Diagram of distribution overlap	Equality or difference — Central tendency	Variability
A	─────────	Same	Same
B	───── ──	Same	Different
C	──── ────	Different	Same
D	─── ─────	Different	Different
E	── ────	Different	Different

cording to the relationship between the differences in central tendency and variability. Five categories are given: *(A)* same central tendency and variability; *(B)* same central tendency but different variability; *(C)* different central tendency but same variability; *(D)* different central tendency and variability, with the *larger* central tendency associated with the *smaller* variability; *(E)* different central tendency and variability, with the *larger* central tendency associated with the *larger* variability.

A discussion of each of the five categories follows. It will be helpful in following the discussion to refer to Table 7.1 from time to time.

Type *A* situation

This situation arises when there is no differential effect of the treatments. But it also occurs when the experiment is not sensitive enough to detect a differential effect. When the differential effects are small and the measurements are variable, a *large number* of *precise* measurements is required to ensure significant results. This is true both of tests of a difference in central tendency and tests of a difference in variability. Thus type *A* results, in which there is no difference in central tendency and no difference in variability, are quite ambiguous. A person cannot tell whether there is *no* differential effect or whether there is a differential effect that the experiment is too insensitive to detect.

Type *B* situation

This situation suggests that forces have pushed or pulled measurements away from (or toward) the center of the distribution, forcing high and low measurements in opposite directions.

Social pressure on a group of persons could cause a closer grouping of purchase prices of new cars than before the attempt to conform. The closer grouping would indicate that people who formerly bought high-priced cars now bought cheaper cars, while those who formerly bought low-priced cars now bought more expensive cars.

There are a number of circumstances where the narrower distribution of a type *B* situation reflects increased proficiency. In learning to operate an elevator, for example, a person will overshoot or undershoot the mark considerably at first but will gradually learn through practice to stop the elevator close to the floor level. A distribution of heights where the elevator stopped at a particular floor becomes narrower as the operator becomes more proficient.

Levine and Murphy (1943) gave students reading matter containing both procommunist and anticommunist views. Students favoring communism remembered the procommunist views better than the anticommunist views, while anticommunist students remembered the anticommunist views better. If it is safe to assume that what they remembered affected their attitudes, the students moved farther away from the neutral point on a scale of attitudes regarding communism, increasing the variability of the attitudes.

Siegel and Tukey (1960) referred to a study by Ellen Tessman in which she showed a movie to two groups of subjects and measured the amount of hostility they saw in the movie. The experimental group consisted of persons who, according to a personality test, had difficulty getting along with people. The control group consisted of persons whose personality test scores indicated that they had good social relations. There was no significant difference in the amount of hostility the two groups saw ($P > .05$), but the experimental group showed more variation ($P < .01$) in the amount of hostility seen. The forces causing persons in the experimental group to have social difficulty may have exaggerated existing perceptual tendencies, causing people ordinarily noticing little hostility to notice less, and people ordinarily noticing a lot of hostility to notice more.

Glick (1959) compared the amount of weight gained by chickens on a standard diet and chickens fed penicillin in addition to the standard ration. Since the average gain in weight was not significantly different ($P > .05$), Glick concluded that the addition of penicillin did not affect weight gain. This conclusion appears to be incorrect because the chickens given penicillin had a greater variability in weight gain ($P < .001$). Instead of having no effect, penicillin has opposing effects: the penicillin decreases the weight gain in chickens that normally would show small gains and increases the weight gain in chickens that normally would make large gains. A chicken farmer who could identify the low-gaining and high-gaining chickens (perhaps on the basis of size at an early age) could supplement the ration of the high-gaining chickens with penicillin and leave penicillin out of the ration of the low-gaining chickens.

Type C situation

The type C overlap suggests that the difference in the effect of the two treatments in the same for high-scorers as for low-scorers. An example of type C overlap is the shift in gross income resulting when every employee of a company is given a $20 Christmas bonus, regardless of his salary.

Mills, Casper, and Bartter (1958) compared the amount of aldosterone (a hormone) in the blood under normal blood pressure with the amount in animals whose blood pressure had been artificially increased by severing the vagus nerve. The increase in blood pressure increased the amount of aldosterone ($P < .001$), but the standard deviation of the amount of the hormone showed no significant change ($P > .05$), suggesting that the change in aldosterone level induced by the increase in blood pressure was independent of the level under normal blood pressure. This suggests the hypothesis that the tissues producing aldosterone normally found in the blood are not the tissues that produce the increment in aldosterone during the increase in blood pressure; however, no organs except the adrenal glands are known to produce aldosterone.

Vowinckel and Orvig (1962) made computations concerning the heat flow in the Arctic Ocean on the assumption that Greenland-Spitzbergen border ice is 50 centimeters thicker than Denmark Strait ice every month of the year even though the ice varies in thickness from month to month. If this assumption is correct, the combination of factors that causes the Greenland-Spitzbergen ice to be thicker than Denmark Strait ice has the same amount of effect on thin ice as on thick ice.

Type *D* situation

In a type *D* situation the differential effect is greater for low-scorers than for high-scorers. One circumstance that can cause this is an upper limit that restricts gains of high-scorers. Such a limit may be inherent in the situation being measured or be the result of limitations in the measuring instrument. As mentioned earlier in regard to the comparison of the two teaching methods, the maximum possible score on a test can restrict the variability of high-scorers.

A similar cause of a type *D* situation is the existence of an upper threshold: a point above which the difference in treatment has no effect. For example, suppose we were to compare the reading speed of a group of readers before and after they were given special instruction in reading without moving the vocal organs. The instruction undoubtedly would speed up the reading of many slow readers who had been saying each word to themselves. On the other hand, fast readers apparently already know how to read without moving their vocal organs or they could not be fast readers; consequently, fast readers will not profit appreciably from the special reading instruction.

The Christmas bonus considered in the type C situation would increase the income *after taxes* in inverse relation to their regular income after taxes, thereby providing a type D situation for income after taxes.

Malpass (1960) compared the motor development of mentally retarded and normal children. The test measured accuracy and quickness of response. The mean score of the normal children was greater ($P < .001$) and the standard deviation smaller ($P < .01$). The upper limits to accuracy and quickness inherent in measurements of distance and time conceivably could account for these results.

Type E situation

The situation is analogous to the type D situation. Instead of an upper limit, however, there is a lower limit that operates in a type E situation.

Sometimes the variability of time required to perform a task is reduced as the average time required is decreased, because long durations of time can be reduced considerably, whereas very short durations of time cannot be reduced much because of the lower limit of time duration. For example, in a study comparing two methods of teaching radio operators to translate code (Keller, 1945), the method with the smaller mean time ($P < .000001$) to pass a qualifying test also had the smaller variability of time ($P < .000001$).

A lower threshold may lead to a type E situation, a lower threshold being a point below which individuals are unaffected by the difference in treatment. For example, consider a comparison of the reading-comprehension scores of a group of beginning readers provided with dictionaries and a group without dictionaries. The beginning readers with poor reading comprehension would be unable to understand the dictionary, and so its availability would not improve their reading comprehension. Readers with a higher comprehension could, however, understand the dictionary and thereby improve their reading-comprehension scores.

A type E situation also arises when a difference in treatment has an effect on scores that is directly related to the rank of the scores. An experimental treatment that resulted in giving all readers in a group the same amount of additional time would increase the amount read more for those who read a large amount during control conditions (the fast readers) than for those who read a smaller amount.

Another example of an effect of a difference in treatment that is probably directly related to the rank of the subjects is an experiment by Mer (1959) on the effect of adding certain sugars to the nutrition of young oat seedlings. The plants given the sugars were longer ($P < .001$) and more variable in length ($P < .01$) than the control plants. The long plants apparently gained more benefit from the sugars than the short ones.

The quantity of diphosphopyridine nucleotide–linked enzymes found in the blood of normal persons and persons with delirium tremens ("DTs") was compared (Allgen et al., 1958). The persons with delirium tremens had larger amounts of the enzymes ($P < .001$), and a greater variability in amount ($P < .001$). A possible lower threshold here is the minimum amount of the enzymes required to stay alive and healthy.

The failure to randomly assign individuals to high-scoring and low-scoring groups

We have considered all five types of relationships between direction of difference in central tendency and direction of difference in variability and possible causes of such relationships. It may appear now that the problem of how to draw inferences from joint analysis of differences in central tendency and variability, if not actually solved, has at least been broken down into workable units. Such confidence is unwarranted, as will be seen when the problem is restated in terms of randomization tests.

The first point to note is that several of the so-called experimental comparisons used as examples do not permit analysis by means of randomization tests because they do not involve random assignment. The comparison of the motor development of mentally retarded and normal children is one of them. The experimenter did not randomly assign children to the "mentally retarded" and "normal" groups and consequently is unable to draw causational inferences about the effects of mental retardation on motor development.

The second point is that even in those experiments which would permit random assignment, like Mer's experiment on the effect of sugars added to the nutrition of young oat seedlings, all that the random assignment permits is statistical inferences about the treatments having a different effect, not about the differential effects being different for high-scoring and low-scoring individuals. Length is not randomly assigned to an oat seedling by the experimenter. Consequently, inferences about the effect of length on response to a treatment must be *nonstatistical*.

There is no way to *completely* overcome the inability to randomly assign certain characteristics, but there are experimental procedures to help solve the problem. Although we cannot, for example, randomly assign persons to "calm" and "anxious" groups, we *can* randomly assign persons to conditions calculated to make them calm or anxious and then compare them on some task. This is by no means the same as comparing already-existing calm and anxious groups, but at least we know something about the conditions that brought on calmness or anxiety in the experiment and therefore have a little better idea of the causal relations than we have with already-existing calm and anxious groups.

The problem of lack of random assignment to high scores and low scores can be handled in a similar fashion. We cannot randomly assign individuals to high and low scores but we *can* randomly assign them to treatments that we expect to make them score high or low, as in assigning seeds to poor or good soil to provide short or long oat seedlings by the time we apply special nutrients to them.

A distribution-free analog of a complex analysis of variance design

Example 7.1

We will now consider how an experiment can be designed to facilitate the making of *statistical* inferences about various treatment effects and *nonstatistical* inferences about the effect of a variable (in this instance, length) which cannot be randomly assigned. Table 7.2 shows how such an experiment might be set up and some hypothetical results.

We will use randomization tests to analyze the data in Table 7.2 in a manner analogous to that employed in analyzing complex analysis of variance tables, wherein the data matrix is subdivided in various ways to permit alternative comparisons within the set of data.

We randomly assign the seeds that produce plants *a* to *l* to two types of soil, six to good soil and six to poor. This is our approximation to random assignment of different lengths to the plants. We measure and record the length of all 12 plants at the end of two weeks. Then we randomly assign three seedlings within each of the soil types to the nourished group, which receives special nutrients. A week later, at the age of three weeks, the length of each of the six nourished and the six unnourished plants is measured and recorded. Then we subtract the length of each plant at two weeks of age from its length at three weeks to determine the growth and record the amount of growth in the table.

It certainly appears that our assignment to different types of soil affected the length of the plants at two weeks in the direction we expected, providing us with short and long plants to be assigned to the

192 Statistical Inference: The Distribution-free Approach

Table 7.2 The length and growth of plants

		Good soil			
		Length (3 weeks)		Change in length (growth)	
Plant	Length (2 weeks)	Unnourished	Nourished	Unnourished	Nourished
a*	5	...	15	...	10
b*	8	...	18	...	10
c	6	11	...	5	...
d	11	20	...	9	...
e	7	16	...	9	...
f*	9	...	20	...	11

		Poor soil			
g	3	5	...	2	...
h*	4	...	9	...	5
i	5	8	...	3	...
j*	6	...	11	...	5
k	4	6	...	2	...
l*	2	...	6	...	4

*Experimental treatment given at 2 weeks.

two nourishment groups. However, if we wanted to, we could test this by using a randomization test on the 12 measurements in the second column, using a difference between poor soil and good soil means as the test statistic. Our initial random assignment, to the two soil types, plus our null hypothesis of no effect of soil type on the length of any of the plants, justifies the use of a randomization test with 12!/6! 6! equally probable divisions of the 12 measurements in the second column into two groups of six measurements.

A different test can be carried out to determine whether the growth of seedlings planted in good soil was affected by the presence or absence of special nourishment. Since we restrict our inferences to the six plants within the good soil group, we ignore the random assignment to soil types and consider only the second random assignment, the assignment to the unnourished and nourished groups. Our null hypothesis is that each of the plants *a, b, c, d, e,* and *f* grew the same amount from the second to the third week as it would have if it had been assigned to the other treatment; that is, the presence or absence of the special nutrient had no effect on the growth of any of those six plants. The values 5, 9, 9, 10, 10, and 11 in the "Change in length" columns are divided in every possible way into three for the unnourished group and three for the nourished group. There are 6!/3! 3! = 20 different ways of dividing the six values. The obtained difference between means in favor of the nourished group is the largest of the 20, and so a one-tailed null hypothesis could be rejected at the .05 level in favor of the alternative hypothesis that the special nutrition favorably affected the growth of at least one of the plants.

The same test could, of course, be carried out to compare the growth of unnourished and nourished seedlings in the *poor* soil. This difference also is significant at the .05 level.

Now in this example the difference between the unnourished and nourished groups is significant within each of the two soil types. With less clear-cut results, however, when we expect the same direction of difference for the plants in both types of soil, we can increase the power of our test by dealing with the difference between the unnourished and nourished plant growth for the two soil types combined. We disregard the random assignment to soil types, and test the null hypothesis that within each of the two soil types the assignment of a plant to the nourished or unnourished condition had no effect on its growth. We do not assume that the growth is the same for these two soil types, which is why we disregard the first random assignment and confine our sampling distribution to equally probable outcomes within each of the two soil types. With our random assignment of plants *within soil types* to unnourished and nourished conditions under our null hypothesis, there are 20 equally probable divisions of the six values within each soil type between the unnourished and nourished conditions. Since the assignment of plants to a nourishment condition within one type of soil is independent of the assignment within the other type of soil, each of the 20 possible arrangements within one soil type is equally probable of being paired with each of the 20 arrangements within the other soil type. Thus, there are 20 × 20 = 400 equally probable arrangements of the 12 growth values where three of the values 5, 9, 9, 10, 10, and 11 are assigned to each of the nourishment conditions and three of the values 2, 3, 2, 5, 5, and 4 are assigned to each of the nourishment conditions. The obtained arrangement, where the highest values for the good soil are in the experimental group and the highest values for the poor soil also are in the experimental group, has the largest difference between means in the predicted direction, and therefore has a probability of 1/400, or .0025.

We cannot perform a randomization test to see whether the shortness or longness of a plant has any influence on the amount the plant grows after it receives the experimental treatment. But we can do the next best thing, which is to see whether assigning a plant to a soil condition likely to make it short or long has such influence. The null hypothesis is that every *nourished* plant grew the same amount from the second to the third week that it would have grown if it were in the other type of soil. For this test we can ignore the *second* random assignment because the inferences are restricted to the plants that were actually assigned to the group receiving special nutrients. That is, we are saying of plants *a, b, f, h, j,* and *l* that the growth of each plant from the second to the third week would have been the same if it had been raised in the other soil and received the special nutrition while it was growing in that soil. We expect long plants to grow more under the nourished condition than short plants, and so we predict that the nourished plants in the good soil will show significantly more growth than those in the poor soil. A randomization test for a difference between the growth of nourished plants in good soil and nourished plants in poor soil would reject the null hypothesis at the .05 level in favor of the alternative hypothesis that at least one of the experimental plants would have grown more in the good soil from the second to the third week than in the poor soil.

Notice that the same kind of test carried out with the *unnourished* plants would reject a null hypothesis about the unnourished plants at the .05 level, based on the numbers in the next to last column, in favor of the alternative hypothesis that at least one of the unnourished plants would have grown more in the good soil from the second to the third week than in the poor soil. Thus, it appears that the amount of growth from the second to the third week, with or without special nutrition, is dependent on the soil in which a plant is raised, the growth being greater in the good soil.

But this does not imply that the plant length itself at two weeks is the influential factor. There may be a number of special effects of the soil that could influence the amount of growth between two and three weeks independently of their influence on the plant length at two weeks: factors that only took effect after that age.

What then has been gained from random assignment to the two soil conditions? One thing we have gained is a narrowing down of possible causes. If we had performed no random assignment to soil at all but just compared the growth of short and long plants from two to three weeks of age, we would not know whether differences are the result of genetic differences in the plants, of soil differences, or of other causes. Our random assignment has at least narrowed down the range of plausible explanations somewhat because we know that the effect occurs even when genetic factors are ruled out.

Furthermore, we can supplement our study by a series of replications in which we have relatively homogeneous soil types and produce the variation in length by means of varying other factors, such as the amount of water given the plant. This could provide, for example, fairly distinctive lengths of plants under two or more watering conditions by the age of two weeks. We would determine, in other words, what effect, if any, the *method* of stunting or stimulating early growth of the plants has on the relative growth of long and short plants between two and three weeks.

In order to see whether the nourishment variable had any effect at all, we statistically compared the changes in length (that is, the growth) of the nourished and the unnourished groups. Mer's study with oat seedlings did not permit him to make a similar comparison because he obtained only one measurement of length from each plant. In Table 7.2 a comparison like his would be that of the nourished and unnourished lengths at three weeks. If we had not measured the plants at two weeks and had to rely on this comparison alone, we would not have significant differences within either the good soil condition or the poor soil condition because of the amount of overlap of the nourished and unnourished distributions within soil conditions.

For both soil conditions we did, however, get significant differences between *growth* of the nourished and unnourished plants; there is no overlap at all between the growth distributions (last two columns) within either soil condition.

Now to determine whether the nourishment variable had any effect at all, it is perfectly valid to test for the significance of a difference in lengths of nourished and unnourished plants at three weeks of age, but a comparison of the *growth* in length is more sensitive. It is more sensitive because the only variability in the numerical values is the variability of growth between two and three weeks. The variability in the distribution of lengths at three weeks of age, on the other hand, has two components: variability of growth plus variability in length at two weeks of age. And, given the reasonable assumption of a positive correlation between the growth and the length at two weeks, the growth spreads the distribution of plant lengths farther apart, making the variability at three weeks greater than at two weeks and greater than the variability of the growth.

On the other hand, suppose there was reason to expect a *negative* correlation between the growth and the length at two weeks. (If the plants attained almost full growth by the age of two weeks, for example, this would be a reasonable expectation.) The addition of small increments to the long plants and large increments to the short ones would tend to *reduce* the variability of the lengths at three weeks. And, if this reduction is large enough, the plant lengths at three weeks could be *less variable* than the growth. It would then be better to use the length at three weeks in testing for a difference between treatments.

Interactions between treatments and measurement magnitudes

The analysis in this chapter of five combinations of direction of difference in central tendency and direction of difference in variability emphasizes the importance of analyzing heterogeneity of variance. The same considerations, however, have implications for the interpretation of interactions, or at least a particular class of interactions.

In the earlier analysis, we discussed the differential effects for high-scoring and low-scoring subjects within a given group under two treatments. Now if we divided our group into two subgroups, a high-scoring and a low-scoring group, and compared the differential treatment effects of the two groups, we would be dealing with an interaction between the measurement magnitudes and the treatments. Figure 7.3 shows graphically the interactions for types *B, D,* and *E* situations and the lack of interaction in the types *A* and *C* situations. This is an alternative way of visualizing the relationships shown in Table 7.1, on page 185.

Interactions between treatments and measurement magnitudes may be rather common. If so, in a number of cases it may not be the apparent variable, but the measurement magnitude associated with it, that interacts with the treatment.

For example, suppose we found an interaction between sex and silent-reading instruction in their effect on reading speed. It is conceivable that the females were faster readers, and therefore profited less from the instruction because they could already read without moving the vocal organs (type *D* interaction).

Also, in the example for a type *E* situation concerning the growth of oat seedlings, one can imagine a comparison of the effects of adding sugar to different types of oat seedlings, one type taking longer to mature than the other. Under these conditions one could get an interaction between the effect of type of oat seedling and nutrition simply because of the interaction between the length of seedling and the treatment.

Not all interactions that look like those diagrammed in Figure 7.3 can be explained in terms of an interaction between treatments and measurement magnitude, but since the diagrammed relationships are

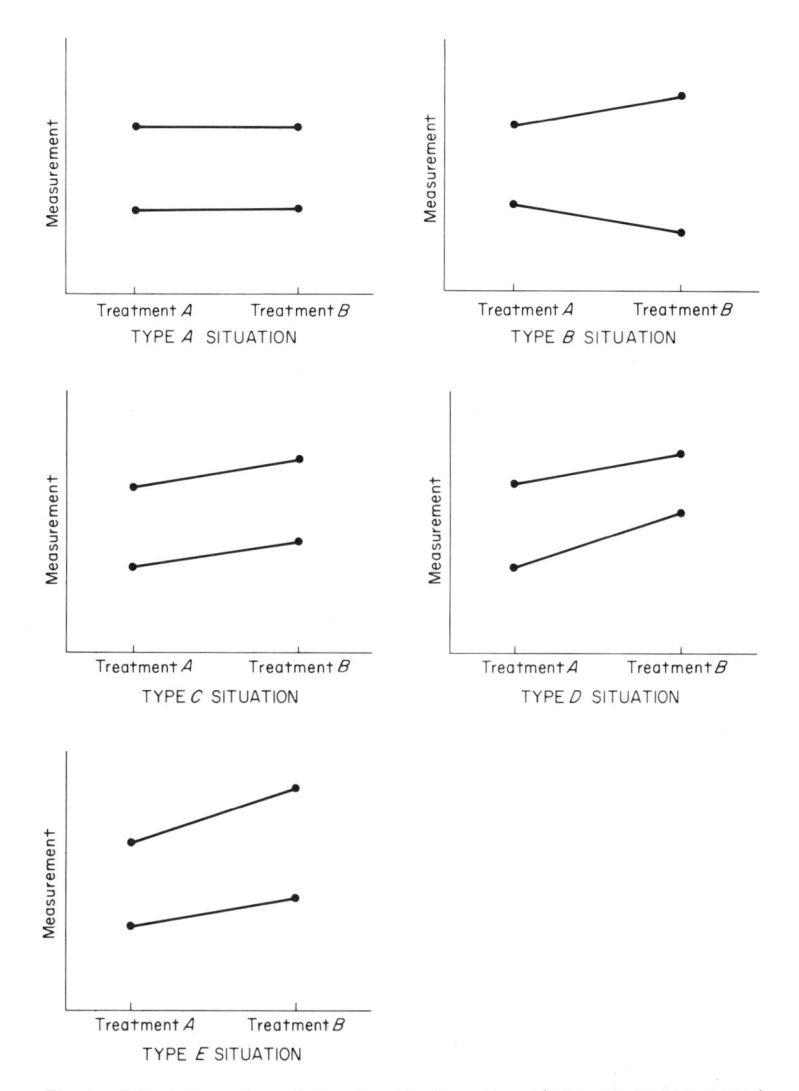

Joint Analysis of Differences in Central Tendency and Variability 197

Figure 7.3 Interactions between treatments and measurement magnitudes. Note: In each graph, the top line connects the means of high-scoring subjects, and the bottom line connects the means of low-scoring subjects.

consistent with such an explanation, it is worth considering this possible interpretation. When there are measurements for the same individuals under both treatments, one can examine the data within the low and the high categories to see whether there is a correlation between the magnitudes within the categories and the differential effect.

The effect of transformations on interactions

In Chapter 5, in the section called "Transformations," it was shown that randomization tests permit statistical inferences about general concepts like "size of a circle" and "speed" because the null hypothesis and the alternative hypothesis concern individual measurements and therefore the statistical inferences have reference to any measurement scale monotonically related to the one employed. Parametric tests, on the other hand, yield inferences about differences between means (unless they are used as approximations to randomization tests, in which case they would yield randomization-test inferences) and therefore permit statistical inferences only about measurements on scales that are linearly related to the one employed. The reason for this restriction on inferences from parametric tests is that the null hypothesis and the alternative hypothesis concern differences between means; the direction of a difference between means (or even the *existence* of a difference between means) is invariant only under linear transformations of measurement scales.

We will now consider the effect of transformations on inferences about interactions. Suppose that we compare the ocular pupil diameter under drugs *A* and *B* and reject the null hypothesis in favor of the alternative hypothesis that at least one individual would have had a larger pupil diameter under *B* than under *A*. Then, as we have said, we could conclude that that individual or those individuals would have had larger pupil *sizes* under *B*, in terms of the area or any measurement monotonically related to diameter, which includes all of the dimensions implied in our concept of the size of a circle.

But suppose we randomly divide the subjects into two groups and hypnotize one group before they are given treatments *A* and *B*. The hypnosis might well cause the pupils to dilate, but we are not interested here in a comparison of the size of the pupils under hypnosis with the size without hypnosis; our interest is in the *difference* in the pupil sizes of the hypnotized persons under the *A* and *B* treatments compared to the difference for the nonhypnotized persons. More specifically, our one-tailed null hypothesis is that none of the subjects would show a greater *B*-minus-*A* differential pupil size if hypnotized than if not hypnotized. That is, none of the nonhypnotized subjects would have shown a larger *B*-minus-*A* differential pupil size if he had been hypnotized and none of the hypnotized subjects would have shown a smaller differential pupil size if he had not been hypnotized.

What generalizations could we make if we measured the pupil diameter and rejected the null hypothesis in favor of concluding that there was an interaction (for at least one subject) between the hypnosis variable and the drugs in their effect on pupil diameter? With our one-tailed null hypothesis stated earlier, this would imply that for at least one of the subjects, the *B*-minus-*A* differential pupil diameter would have been larger for the hypnotized condition than for the nonhypnotized condition. Does this also imply that the subject would have a larger *B*-minus-*A* differential pupil *area* for the hypnotized condition? No, not necessarily, because a comparison of differential pupil areas is not a comparison of individual measurements but a comparison of differences between individual measurements; and the direction of the difference between two differences between individual measurements is invariant only under a linear transformation of the measurements, just as means are. For example, the hypnosis may provide the larger *B*-minus-*A* differential pupil diameter but the *smaller* differential pupil area, for an individual. Therefore with randomization tests one could not test for an interaction of two variables on pupil *size,* but only on certain dimensions of size that are linearly related. Instead then of being able to generalize to any of the set of monotonically related measurement scales as was possible in simple comparisons of two treatments, interaction conclusions are restricted to the narrower class of linearly related measurement scales.

It is of course true that parametric tests for interactions, which yield inferences about the interaction between two variables in their effect on a mean, would involve differences between means, and so their inferences also would be restricted to the class of measurements provided by linearly related measurement scales.

If the differential pupil diameter of a subject was larger when hypnotized than when not hypnotized, then the differential pupil radius and the differential pupil circumference also must have been larger, but not necessarily the differential pupil area. This suggests that there may be empirical reasons for expecting an interaction for one measurement scale but not for a nonlinear transformation of the scale. We will now consider this possibility.

For instance, suppose that we divided some bean seedlings into two classes, long seedlings and short seedlings, and obtained the means plotted in Figure 7.4, at three days and at eight days of age. We note that the long seedlings grew more than the short ones from the third to the eighth day. Thus we have nonparallel lines in the graph, indicating an interaction. However, we can see that although the short plants and the long plants gained different amounts of length, the increase was proportional to the mean length on the third day; for each group of plants the increase in the mean is three times the original mean length.

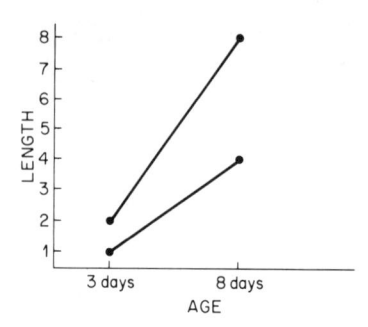

200 Statistical Inference: The Distribution-free Approach

Figure 7.4 Interaction between age and length for bean seedlings.

Now if Figure 7.4 were a plot of four *measurements* instead of four *means*, we would know that a plot of the logarithms of the measurements would have parallel lines, showing a lack of interaction. We cannot, however, obtain the means of logarithms of measurements by taking the logarithms of the means of the measurements, and so we cannot determine the means of the logarithms of the measurements from the information in Figure 7.4. Nevertheless, if the individual measurements were accessible to us and we computed the mean of the logarithms for each of the four groups of measurements, they would probably yield a plot very similar to Figure 7.5, in which there is no interaction. (The values in Figure 7.5 are the logarithms of the values plotted in Figure 7.4.) The interaction between length and age shown in Figure 7.4 indicates that the long plants increased in length from the third day to the eighth day more than the short plants did. And the lack of interaction in Figure 7.5 between age and the logarithms of length indicates that the *proportional* increase in length was the same for tall and short plants.

Geometric rates of increase are so common in biology that it might seem reasonable to expect the amount of increase in the size of the bean seedlings from three days to eight days of age to be proportional to the size at three days. But we are not simply concerned with proportional changes in *size* but with proportional changes in *length*. Should we expect the increase in length to be proportional to the initial length or expect the increase in *volume* to be proportional to the initial *volume*? Since length and volume are not linearly related, there cannot be proportional increases in both; if one increases proportionally to the initial magnitude, the other cannot. Or, as the surface area is responsible for collecting the food for the plant, perhaps the change in the *volume* over a period of time is proportional to the initial surface area. Even this concept oversimplifies the complexities of the growth process, though, be-

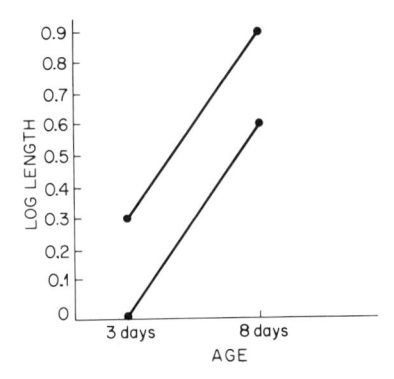

Figure 7.5 Absence of interaction between age and logarithms of length for bean seedlings.

cause we are referring to an "instantaneous" rate of change, whereas over any period of time the surface area is increasing at a slower rate than the volume.

To know what sort of interaction to expect or how to interpret an obtained interaction can, then, require rather detailed consideration of the empirical processes through which the independent variables are assumed to exert their effects.

References

Allgen, Lars-Goran, Sander Izikowitz, Britta Nauckhoff, Inga-Britt Ordell, and Inna Salum. 1958. Serum diphosphopyridine nucleotide-linked enzymes in delirium tremens and allied conditions. *Science,* **128,** 304-305.

Almy, M. 1955. *Child development.* New York: Holt, Rinehart and Winston, Inc.

Andrewartha, H. G. 1961. *Introduction to the study of animal populations.* Chicago: The University of Chicago Press.

Blanchard, J. 1918. The brightness sensibility of the retina. *Physical Review,* **11,** 81-99.

Boring, E. G. 1942. *Sensation and perception in the history of experimental psychology.* New York: Appleton-Century-Crofts, Inc.

Campbell, D. T. 1957. Factors relevant to the validity of experiments in social settings. *Psychological Bulletin,* **54,** 297-312.

Campbell, D. T., and J. C. Stanley. 1966. *Experimental and quasi-experimental designs for research.* Chicago: Rand McNally and Company.

Dukes, W. F. 1965. $N = 1$. *Psychological Bulletin,* **64,** 74-79.

Eden, T., and F. Yates. 1933. On the validity of Fisher's z-test when applied to an actual sample of nonnormal data. *Journal of Agricultural Science,* **23,** 6-16.

Edgington, E. S. 1960a. Contradictory conclusions from two speed of performance measures. *Psychological Bulletin,* **57,** 315-317.

———— 1960b. Nonlinearly related measurement scales. *Journal of Psychology,* **50,** 399-402.

———— 1961a. Probability table for number of runs of signs of first differences in ordered series. *Journal of the American Statistical Association,* **56,** 156-159.

———— 1961b. A statistical test for cyclical trends, with application to Weber's law. *American Journal of Psychology,* **74** (no. 4), 630-632.

———— 1963. Joint analysis of differences in central tendency and variability. *American Statistician,* **17,** 28-30.

———— 1966a. Statistical inference and nonrandom samples. *Psychological Bulletin,* **66,** 485-487.

———— 1966b. Implications of symmetry about a regression line. *Perceptual and Motor Skills,* **23,** 321-322.

———— 1967. Statistical inference from $N = 1$ experiments. *Journal of Psychology,* **65,** 195-199.

———— 1969. Approximate randomization tests. *Journal of Psychology,* **72,** 143-149.

Evans, F. C. 1949. A population study of the house mouse (*Mus musculus*) following a period of local abundance. *Journal of Mammalogy,* **30,** 351-363.

Fisher, R. A. 1935. *The design of experiments.* Edinburgh. Oliver & Boyd Ltd.

Glick, Bruce. 1959. The experimental production of the stress picture with cortisone and the effect of penicillin in young chickens. *Ohio Journal of Science,* **59,** 81-86.

Holway, A. H., and C. C. Pratt. 1936. The Weber-ratio for intensity discrimination. *Psychological Review,* **43,** 322-340.

Keller, F. S. 1945. *The radio code project: Final report of Project SC-88 OSRD Report 5379.* The Psychological Corporation.

Kempthorne, O. 1952. *The design and analysis of experiments.* New York: John Wiley & Sons, Inc.

———— 1955. The randomization theory of statistical inference. *Journal of the American Statistical Association,* **50,** 946-967.

Kendall, M. G. 1943. *The advanced theory of statistics.* Vol. 1. London: Charles Griffin & Company, Ltd.

Levine, J. M., and G. Murphy. 1943. The learning and forgetting of controversial material. *Journal of Abnormal and Social Psychology,* **38,** 507-517.

Lindquist, E. F. 1953. *Design and analysis of experiments in psychology and education.* Boston: Riverside Editions, Houghton Mifflin Company.

Malpass, Leslie F. 1960. Motor proficiency in institutionalized and non-institutionalized retarded children and normal children. *American Journal of Mental Deficiency,* **64,** 1012-1015.

Mer, C. L. 1959. The analysis of correlative growth in the etiolated oat seedling in relation to carbon dioxide and nutrient supply. *Annals of Botany,* **23** (no. 89), 177-194.

Meredith, W. M. 1967. *Basic mathematical and statistical tables for psychology and education.* New York: McGraw-Hill Book Company.

Mills, Ivor H., Alfred Casper, and Frederic C. Bartter. 1958. On the role of the vagus in control of aldosterone secretion. *Science,* **128,** 1140-1141.

Milne, A. 1943. The comparison of sheep tick populations (*Ixodes ricinus* L.). *Annals of Applied Biology,* **30,** 240-250.

Pitman, E. J. G. 1937. Significance tests which may be applied to samples from any populations: III. The analysis of variance test. *Biometrika,* **29,** 322-335.

Salt, G., F. S. J. Hollick, F. Raw, and M. V. Brian. 1948. The arthropod population of pasture soil. *Journal of Animal Ecology,* **17,** 139-150.

Sidman, M. 1960. *Tactics of scientific research; evaluating experimental data in psychology.* New York: Basic Books, Inc., Publishers.

Siegel, S. 1956. *Nonparametric statistics for the behavioral sciences.* New York: McGraw-Hill Book Company. Pp. 206-207.

Siegel, S., and John W. Tukey. 1960. A nonparametric sum of ranks procedure for relative spread in unpaired samples. *Journal of the American Statistical Association,* **55,** 429-445.

Silvey, S. D. 1954. The asymptotic distributions of statistics arising in certain nonparametric tests. *Proceedings of the Glasgow Mathematics Association,* **2,** 47-51.

Skinner, B. F. 1938. *The behavior of organisms: an experimental analysis.* New York: Appleton-Century-Crofts, Inc.

Stevens, S. S. 1951. Mathematics, measurement, and psychophysics. In S. S. Stevens (Ed.), *Handbook of experimental psychology.* New York: John Wiley & Sons, Inc. Pp. 1-49.

Vowinckel, Eberhard, and Svenn Orvig. 1962. Water balance and heat flow of the Arctic Ocean, *Arctic,* **15** (no. 3), 205-223.

Wald, A., and J. Wolfowitz. 1944. Statistical tests based on permutations of the observations. *Annals of Mathematical Statistics,* **15,** 358-372.

Walker, H. M., and J. Lev. 1953. *Statistical inference.* New York: Holt, Rinehart and Winston, Inc.

Wallis, W. A., and H. V. Roberts. 1956. *Statistics: a new approach.* Brooklyn: Free Press.

Welch, B. L. 1937. On the z-test in randomized blocks and Latin squares. *Biometrika,* **29,** 21-52.

Winer, B. J. 1962. *Statistical principles in experimental design.* New York: McGraw-Hill Book Company.

Index

Allgen, L., 190
Almy, M., 83
Analysis of variance:
 as approximation to randomization test, 162-164
 complex, randomization test analog of, 191-196
 unequal sample sizes in, 163-164
Andrewartha, H. G., 42, 43
Approximate randomization tests, 152-155
 power of, 153-155
 validity of, 152-153
Approximate sampling distribution, 152
Area-to-diameter transformation, 158-160
 (See also Transformations of measurements, nonlinear)
Assumption of normality (see Normal distribution assumption)

Bartter, F. C., 188
Binomial distribution, 76-77
Bivariate normal distribution assumption, 71, 81-83
Blanchard, J., 179
Blood pressure, effect of aldosterone on, 188
Boring, E. G., 179

Campbell, D. T., 97
Cartilage cells, 148
Casper, A., 188
Cells:
 clumping of, 149
 nose cartilage, 148
Central limit theorem, 73-75, 81-82
Central tendency and variability differences, 184-190
Centroids, sample, 81-82
Change measurements in longitudinal experiments, 144-146
Chickens with penicillin added to their diet, 187
Classical probability, 8-9
Clumping of cells, 149
Code, teaching translation of, 189
Coin tossing, 13-15
Combinations, formula for, 19
Communism, attitudes regarding, 187
Confidence and probability, 53

Confidence intervals:
 based on linear regression, 71-73
 distribution-free: "at least" levels of confidence, 53-55
 for a difference between means, 61-63
 for a difference between totals, 62-63
 for an individual, 63-65
 for a mean, 59-61
 for a median, 55-56
 for a proportion, 57-59
 for a total, 59-61
 exclusive and inclusive, 54
 normal curve: for a difference between totals, 66-68
 for an individual, 68-71
 for the mean of a different sample, 68-71
 for a proportion, 66
 purposes of, 50-53
Contingency chi-square test, 125-128
 (See also Fisher's exact test; Randomization tests, for contingency)
Correlated measurements, 146-150
 (See also Experimental independence)
Correlation:
 between height and weight, 82-83
 product-moment, 166-167
 randomization test for, 115-118
 rank-order, 128-130
 Spearman's rank, 128-130

D'Alembert, J., 13
Delirium tremens, enzymes in persons with, 190
Denmark Strait ice thickness, 188
Diameter-to-area transformation, 158-160
 (See also Transformations of measurements, nonlinear)
Dice rolling, 11-13
Difference between means, distribution-free confidence interval for, 61-63
Difference between totals:
 distribution-free confidence interval for, 62-63
 normal curve confidence interval for, 66-68
 equal population sizes, 66-67
 unequal population sizes, 67-68
Differences in central tendency and variability, 184-190

Differential effect of treatments, 183–184
Differential scores in randomization test for interaction, 119
Distribution-free approach, definition of, 2
Distribution-free tests (see Randomization tests)
Dukes, W. F., 135

Eden, T., 161
Edgington, E. S., 97, 135, 152, 156, 158, 175, 179, 185
Enzymes in normal persons and persons with delirium tremens, 190
Equal-sized rejection regions for two-tailed tests, 103–105
Equally likely events, 9–11
  based on insufficient reason, 10
  based on sufficient reason, 10
Estimation in psychological testing, 90–91
Evans, F. C., 43
Experimental independence, 150–151
  (See also Statistical independence)

Fisher, R. A., 94, 161
Fisher's exact test, 113–115
  (See also Contingency chi-square test; Randomization tests, for contingency)

Gambling:
  coin tossing, 13–15
  dice rolling, 11–13
  law of averages, 15
  martingales, 15
  maturing of the chances, 15–16
General laws, 170–171
Generality of randomization test statistical inferences, 160–161
Generalization, nonstatistical (See Inferences, nonstatistical)
Glick, B., 187
Greenland-Spitzbergen ice thickness, 188

Height, correlation of weight with, 82–83
Heterogeneity:
  of cell size, 148
  of subjects, arguments for, 141–142
High-scoring and low-scoring subjects, 184–191
Holway, A. H., 179

Homogeneity of subjects, arguments for, 141–142
Hormone aldosterone, effect on blood pressure of, 188

Ice, thickness of, 188
Independence:
  experimental, 150–151
  statistical, 146–150
Inferences:
  nonstatistical: compared with statistical, 89–90
    experimental design to facilitate making of, 140–142, 171–172
    when population contains unmeasurable individuals, 87–89
  statistical: compared with nonstatistical, 89–90
    generality of, for randomization tests and normal curve tests, 159–161, 198–200
    in one-subject experiments, 135–140
    restricted to measurable individuals, 87–89
Insufficient reason, 10
Interaction:
  effect of nonlinear transformations on, 198–200
  normal curve test for, 119–120
  randomization test for, 118–120
  specificity of statistical inferences about, 198–200
  between treatments and measurement magnitudes, 196–197
Invariance of the mean (see Mean, invariance of)
Invariance of randomization test probabilities:
  under linear transformations of measurements, 166–167
  under ordinal transformations of test statistics (see Test statistics)
Isolation of treatments, temporal, 108–109, 139

Keller, F. S., 189
Kempthorne, O., 94, 97, 161
Kendall, M. G., 31n.

Law of averages, 15
Laws, general, 170–171

Learning experiment with rats, 146–147
Lev, J., 73
Levine, J. M., 187
Lindquist, E. F., 139
Linear regression, confidence intervals based on, 71–73
(See also Regression)
Long-run relative frequency probability, 7–8
Longitudinal experiments, 142–146
(See also One-subject experiments, statistical inferences for)

Malpass, L. F., 189
Mann-Whitney U test, 123–124
Martingales, 15
Maturing of the chances, 15–16
Maximum heterogeneity of subjects, 141–142
Maximum homogeneity of subjects, 141–142
Mean:
 distribution-free confidence interval for, 59–61
 compared to interval for the median, 65–66
 invariance of: lack of: under diameter-to-area transformations, 158–161
  under reciprocal transformations, 156–158, 160
  under linear transformations, 79–81
  (See also Transformations of measurements, nonlinear)
 standard error of, 31
Means, distribution-free confidence interval for difference between, 61–63
Measurements:
 of change in longitudinal experiments, 144–146
 correlated, 146–150
 (See also Experimental independence)
 logarithmic transformations of, 80, 145, 165–166, 176
 reciprocal transformations of, 156–158, 160
 reliability of, 84–87
 repeated, and statistical independence, 146–150
 tied (see Tied measurements)
 validity of, 83–84
 (See also Test statistics; Transformations of measurements, nonlinear)

Median, distribution-free confidence interval for, 55–56
 compared to interval for mean, 65–66
Mentally retarded and normal children, accuracy and speed of, 189
Mer, C. L., 189
Meredith, W. M., 56, 124
Mills, I. H., 188
Milne, A., 43
Movie, amount of hostility seen in, 187
Murphy, E., 187

Nonmonotonic trends, test for, 178–182
Nonparametric tests (see Randomization tests; Rank-order tests)
Nonrandom samples in psychological experiments:
 adequacy of, 97–98
 prevalence of, 96–97
Nonstatistical inferences (see Inferences, nonstatistical)
Normal curve tests:
 as approximations to randomization tests, 161–167
 analysis of variance, 162–164
 product-moment correlation, 166–167
 t test, 161
  for paired comparisons, 164–166
 for interaction, 119–120
 specificity of statistical inferences compared to randomization tests, 158–161
Normal distribution assumption:
 binomial distribution argument, 76–77
 bivariate, 71, 81–83
 central limit theorem argument, 73–75
 nonmathematical arguments, 77–79
Nose, cartilage cells from, 148
Null hypothesis:
 one-tailed and two-tailed, 96
 plausibility of, 167–168
 for tests of interaction, 119–120
 for randomization tests, 95–96
 in terms of individuals, not test statistics, 120–122

Oat seedlings, sugar added to nutrition of, 189–190
One-subject experiments, statistical inferences for, 135–140
(See also Longitudinal experiments)

Index

One-tailed test, sampling distribution for, 137–138
   for analysis of variance, 163–164
One-tailed and two-tailed probabilities:
   in analysis of variance: with equal sample sizes, 163
      with unequal sample sizes, 163–164
   for the same test statistic, 137–138
Operational definition of probability, 9–11
Ordinally related test statistics (*see* Test statistics)
Orvig, S., 188

Parametric tests (*see* Normal curve tests)
Penicillin, its effect on chickens, 187
Perfect-experimental-control regression lines, 175
Permutation tests (*see* Randomization tests)
Pitman, E. J. G., 161
Plants:
   in good and poor soil, 191–196
   sugar added to nutrition of, 189–190
Power:
   of approximate randomization tests, 153–155
   of randomization tests, 93
Pratt, C. C., 179
Probability:
   classical, 8–9
   and confidence, 53
   long-run relative frequency, 7–8
   operational definition of, 9–11
   and significance, 95
   subjective, 7
Probability table for a test of nonmonotonic trends, 181
Probability tree, 11
Product-moment correlation, as approximation to a randomization test, 166–167
Proportion:
   distribution-free confidence interval for, 57–59
   normal curve confidence interval for, 66
Psychological testing:
   applied, 90–91
   theory, 90–91

Radio operators, learning of code by, 189

Random assignment, 93–94
Randomization test probabilities, invariance of: under linear transformations of measurements, 166–167
   under ordinal transformations of test statistics (*see* Test statistics)
Randomization tests:
   approximate, 152–155
   for contingency, 109–115
      equivalence to Fisher's exact test, 114–115
      (*See also* Contingency chi-square test; Fisher's exact test)
   for correlation, 115–118
   for a difference between independent samples, 98–103
   for a difference between regression lines, 177–178
   distribution-free analog of a complex analysis of variance design, 191–196
   generality of statistical inferences compared to normal curve tests, 159–161
   for interaction, 118–120, 198–200
      differential scores in, 119
      implausibility of null hypothesis for, 119–120
   invariance of probabilities: under linear transformations of measurements, 166–167
      under ordinal transformations of test statistics (*see* Test statistics)
   null hypothesis for, 95–96
      in terms of individuals, not test statistics, 120–122
   for paired comparisons, 105–109
   power of, 93
Rank-order tests:
   contingency chi-square test, 125–128
      (*See also* Fisher's exact test; Randomization tests, for contingency)
   for correlation, 128–130
   for independent samples, 123–124
   Mann-Whitney *U* test, 123–124
   for nonmonotonic trends, 178–182
   for paired comparisons, 124–125
   as randomization tests for ranks, 123
   Wilcoxon matched-pairs signed-ranks test, 124–125
Rats, learning experiment using, 146–147

Reading:
  effect of giving extra time for, 189
  effect of providing readers with dictionaries, 189
  effect of special silent-reading instruction on, 188
Reciprocally related measurements, 156–158, 160
Regression, comparison of experimental and nonexperimental, 169–170, 173–174
Regression lines:
  fitting of, 172–173
  perfect-experimental-control, 175
  randomization test for a difference between, 177–178
  simultaneous-measurement, 173–175
  symmetry about, 176–177
    logarithmic transformations to provide, 176
Rejection regions for two-tailed tests, equal-sized, 103–105
Reliability of measurements, 84–87
Repeated measurements and statistical independence, 146–150
Retarded and normal children, accuracy and speed of, 189
Roberts, H. V., 23, 24, 37, 38–41
Runs of plus and minus signs, test for, 178–182

Salt, G., 43
Sampling:
  of animal populations, 42–44
    capture-recapture, 43–44
    random, 42–43
    release-capture, 43
  area, 45
  bias from sample dropouts, 45–47
  cluster, 33–37
    and statistical independence, 150
    stratified, 37
  efficiency of, 24–25
  haphazard, 47
  indirect, 38–40, 45
  random, 21–23
  reasons for, 23–24
  sequential, 26–27
  simple random, 26
  stratified, 27–33
    nonproportional, 31–33
    proportional, 32
  of subpopulations, 40–42
Sampling distribution for one-tailed test, 137–138, 163–164
Sequential effects, 108–109, 139
Sidman, M., 135

Siegel, S., 134, 187
Significance and probability, 95
Significance tests (see Normal curve tests; Randomization tests; Rank-order tests)
Silvey, S. D., 161
Simultaneous-measurement regression lines, 173–175
Single-subject experiments (see One-subject experiments)
Skinner, B. F., 135
Spacing of the independent variable, 173
Spearman's rank correlation, 128–130
Spearman's rho, 129, 133–135
  with many ties, 133–135
Specificity of normal curve statistical inferences, 158–161, 199
Specificity of randomization test statistical inferences, 199
Speed of performance measures, 156–158
Standard error of estimate, 72–73
Standard error of the mean, 31
Stanley, J. C., 97
Statistical independence, 146–150
  (See also Experimental independence)
Statistical inferences (see Inferences, statistical)
Stevens, S. S., 79, 155–156
Subjective probability, 7
Sufficient reason, 10

t test:
  as approximation to a randomization test, 161
  for paired comparisons, as approximation to a randomization test, 164–166
Temporal isolation of treatments, 108–109, 139
Tessman, E., 187
Test for nonmonotonic trends, 178–182
Test statistics, equivalence of ordinally related: for correlation, 129–130
  for a difference between independent samples, 102–103, 123–124, 163
  for paired comparisons, 106–107, 165
  for a test of contingency, 111
Testing, psychological (see Psychological testing)
Tests of significance (see Normal curve tests; Randomization tests; Rank-order tests)

Tied measurements, possibility of:
    implications for distribution-free confidence intervals, 54–55
    implications for randomization tests, 95
Tied ranks, 130–135
  Spearman's rho with many, 133–135
Tissue slices, 148–149
Total, distribution-free confidence interval for, 59–61
Transformations of measurements,
    nonlinear: effect on interactions, 198–200
  exponential: in correlation, 83
  diameter-to-area, 158–160, 173
  logarithmic: implications for confidence intervals, 80
    in longitudinal experiments, 145
    in paired comparisons, 165–166
    for producing symmetry about a regression line, 176
  reciprocal, 156–158, 160
Tukey, J. W., 187
Two-tailed tests, procedure providing equal-sized rejection regions for, 103–105

Vaccinated animals, 151
Validity:
  of approximate randomization tests, 152–153
  of measurements, 83–84
Vowinckel, E., 188

Wald, A., 161
Walker, H. M., 73
Wallis, W. A., 23, 24, 37–41
Weber's law, 179–180
Weight, correlation of height with, 82–83
Welch, B. L., 161
Wilcoxon matched-pairs signed-ranks test, 124–125
Winer, B. J., 139
Wolfowitz, J., 161

Yates, F., 161

This book was set in News Gothic by John C. Meyer & Son, and printed on permanent paper and bound by The Maple Press Company. The designer was Marsha Cohen; the drawings were done by J. & R. Technical Services, Inc. The editors were Walter Maytham and George A. Rowland. Stuart Levine supervised the production.

STATISTICAL TESTS* FOR:

TYPE OF TEST	Difference between independent samples	Paired comparisons	Contingency	Correlation	Interaction	Difference between regression lines
Randomization†	pp. 98-105, 191-197	pp. 105-109	pp. 109-115	pp. 115-118	pp. 118-120	pp. 177-178
Rank-order	Mann-Whitney U test, pp. 123-124	Wilcoxon matched-pairs signed-ranks test, pp. 124-125	Chi-square, pp. 125-128	Spearman's, pp. 128-130 Test for nonmonotonic trends, pp. 178-182		
Normal curve	Analysis of Variance, pp. 162-164	t test for paired comparisons, pp. 164-166		Product-moment, pp. 166-167		

*For statistical testing with only one experimental subject, see "One-subject experiments," pp. 135-140.
†See also "Approximate randomization tests," pp. 152-155.